PARIS JOURNAL

1956-1965

PARIS JOURNAL

VOLUME TWO
1956–1965

JANET FLANNER (GENÊT)

EDITED BY WILLIAM SHAWN

A Harvest / HBJ Book

HARCOURT BRACE JOVANOVICH, PUBLISHERS

San Diego New York London

Paris Journal, Volume 1 1944–1955 and *Paris Journal, Volume 2 1956–1964* were previously published together as *Paris Journal, Volume 1 1944–1965.*

Library of Congress Cataloging-in-Publication Data

Flanner, Janet, 1892-
Paris journal.

"A Harvest/HBJ book."
Includes indexes.
Contents: v. 1. 1944-1955—v. 2. 1956-1964—v. 3. 1965-1970.
1. Paris (France)—History—1944- . 2. France—Politics and government—1945- . 3. Paris (France)—Intellectual life—20th century. I. Shawn, William. II. Title.
DC737.F55 1988 944′.361082 88-2300
ISBN 0-15-670948-1 (v. 1)
ISBN 0-15-670949-X (v. 2)
ISBN 0-15-670952-X (v. 3)

Printed in the United States of America

First edition

A B C D E

To my editor, William Shawn,

with admiration, gratitude,

and more than affection

1956

January 5

The Tour Eiffel caught fire the day after the elections. It was something that no one in France expected and that could hardly be believed when it happened—and it was a perfect symbol of the election results. Nobody had been able to imagine anything worse than that the new Assembly should turn out to resemble the old one. Neither politicians nor voters had any advance notion of the curious monster they were creating between them, with its body paralyzed in the middle and with swollen, extraneous wings on its left and right. It was the wild success of the Poujadists that upset all predictions and calculations and has left in its destructive wake nothing but explanations as strange as the victory itself. According to *Figaro,* it is now clear—too late—that the Poujadists are the myriads of little forgotten and discontented men, anti-parliamentarian and leaderless, who rose with a shout in 1951 to follow the nobilities of General de Gaulle and this year settled for the fisticuffs of the stationery-shop keeper of St.-Céré, who thus inherited de Gaulle's leadership in an illegitimacy of descent such as no previous political campaign has ever seen. There was certainly nothing secretive or hidden in the Poujadists' campaign, animated as it was by shouting hecklers at rival meetings, yells, fights, threats, and a violence of language and sentiment that echoed in every corner of France. Maybe all this deafened the Institut Français d'Opinion Publique, the French Gallup Poll, which heard nothing at all to indicate that Poujade's men would come out of their fights with fifty-one Assembly seats. It was predicted that the Communists would win a couple of dozen additional seats, but no one dreamed they would win fifty-two new ones, which makes them the dominant party, with a hundred and forty-five deputies, and, furthermore, the only established party in all France to win new seats, every other party having lost in strength. Never before in the electoral history of the four Republics have all the parliamentary parties slid downhill together this way, in a crumbling decline set going by a kick from the discontented voters, and never before have the only parties to pick up new seats been two anti-parliamentary parties.

On Tuesday night, with the returns mostly in, *Le Monde* went straight to the heart of the matter. "The problem of tomorrow" it stated, "is to ascertain whether, outside of the Communists and Poujadists, a government majority can be located." On Wednesday, two days after the election, the Communist paper *Humanité* suddenly ran an enormous headline: "VIVE LE FRONT POPULAIRE!" *Huma's* idea of the Popular Front today is a kind of Communist houseparty where everyone is welcome to stop in and make common cause against the reactionaries, and since both the Socialists and Mendès-France's adherents have refused the invitation, it now appears that the guests for whom *Humanité* is hanging out the latchstring must be the reactionary Poujadists themselves, though they are considered by everybody else the fiercest and most ignorant reactionaries now going. Nobody in the center is blind to the fact that the two parties, rigged up in an unnatural plurality, would be a terrible combination. The Faurist majority no longer exists, and the strict Mendèsists, too, are a minority. The number of Mendèsist deputies elected is probably around a hundred and fifty, and of Faurists perhaps two hundred; nobody can tell exactly, because a baker's dozen or so on each side are affiliates who defy positive categorizing until tried and proved. The only optimistic prophecy given out so far on the future of this new Parliament is that it cannot last more than six months before being dissolved.

The thing that is clearest now is the picture of the voters—of their energy and determination to express themselves, and their definite frame of mind, which is largely discontent with what they have been receiving from the government as it has been set up recently, and also over the past ten years. This is considered the only reading possible for the dwindling number of seats they gave the old-line, established French parties. It is the key reading for the big gain they gave the Communists and for the landslide sendoff they proffered the Poujadists—votes representing, to an unidentifiable degree, social protest, dissatisfaction, and hopes for change, rather than any recent flood of dialectical conversions or any great historical conviction as to Poujade's odd notion of reviving the Etats Généraux of 1789 (which, he would have discovered if he had read a little farther in Larousse, resulted in a governmental paralysis that led to the creation of a National Assembly, which is what France has now and which has long been paralyzed in its turn). This time, the voters were voting for what they thought they and France ought to have now, not someday, and for the politicians or groups most likely to be able to provide at least part of it—and, above all, for men who would get something, amost anything, done.

In an expectedly heavy turnout, over eighty per cent of the voting population went to the polls—a figure that makes American election intensities look puny. There has not been a winter national election in France since 1876, winter being considered a poor time to get people out to vote. And the day after New Year's, the biggest annual family holiday, is the worst day for an election in the whole year. Parisians who had gone off on the traditional holiday cut their pleasures short in the civic resolve, rare in any democracy, to get back and do their duty. The state railways assisted, with more than a hundred extra so-called voters' trains coming into the Gare de Lyon alone on Sunday and Monday mornings from the South—especially from the Riviera. Beginning on Monday, motor routes leading to Paris were reported thick with family cars of all grades, bringing the adults—generally accompanied by country baskets of Brussels sprouts and pale winter butter, about the only farm produce available just now—to the *urnes*. The importance of this big turnout was that it constituted a grave warning. One of the most bitterly popular slogans was *"Sortez les sortants!"* ("Throw out the outgoing deputies!"), and a hundred and forty-six of them were so thrown into political oblivion. The Gaullist Social Republican Party was almost wiped out, Bidault's M.R.P.s took a bad drubbing, and the theory and functioning of Parliament itself once more suffered a hard beating. However, according to Wednesday night's *Le Monde,* Parliament will have its new chance in the hands of the Socialists and of Mendès-France, on both of whom the majority and the government of tomorrow will depend. That can well be a lot better luck than Parliament and a great many voters deserve.

January 17

The immortal Mistinguett has finally died, at the age of eighty-two. Her obituaries were as spectacular as the finales of her prewar Folies-Bergère revues. Academician Jean Cocteau contributed one that scintillated with melancholy, like paillettes shining on her dramatic black stage gowns. "Her small face," he declared, "seemed like that of a little girl, sculptured as if by the blows of childhood. Her voice, slightly off-key, was that of the Parisian street hawkers—the husky, trailing voice of the Paris people. She was of the animal race that owes nothing to intellectualism. She incarnated herself. She flattered a French patriotism that was not shameful. It is normal that now she should crumble, like the other caryatids of that great and marvellous epoch that was ours."

Known to all France as "Miss"—pronounced "Mees"—she was familiarly tutoied by her affectionate audiences, who would shout from the back of the Casino, *"T'es belle, toi!"* It was glory. She started her career in 1890, and for forty-nine years her Parisian genius was her figure, her optimistic *brio,* her penetrating, touching voice, her thirty-two white indestructible teeth (she had them all till the end), and her million-dollar legs (always prudently covered with wool fleshings, against drafts, beneath her silk tights). In 1911, she entered the Folies-Bergère and there met young Maurice Chevalier, whose talent she perceived with her heart—as one of her obituaries stated—as well as with her acute theatre sense, and they became the great paired stars. In such songs as "C'est Mon Homme" and "J'en Ai Marre," she carried the torch of the *filles de Paris*—of the *gigolette* in the Rue de Lappe, of the sidewalk *gonzesse* and the apache *gommeuse* of Montmartre. She was not held to have advanced French culture, but between the wars she helped make Paris famous again. She was a flower seller as a little girl. Her parents were mattress makers. She was born Jeanne Bourgeois. Around 1897, she concocted her stage name of Mistinguett in imitation of a popular song called "La Vertinguette." Mistinguett loved her fame, the footlights, and her money—she earned a fortune—and was never tempted by marriage. By a liaison with a Brazilian admirer, she had a son, Leopoldo de Lima et Silva, who is now a well-known Paris doctor and was her physician on her deathbed. Colette once said of Mistinguett, "She is a national property."

February 1

The investiture of Parliament yesterday was marked by an extreme lack of enthusiasm and applause. All that anybody really knows about it and its Front Républicain government (Socialist Premier Guy Mollet and Radical Pierre Mendès-France, Minister without Portfolio, are the two men who count) is that either it may or may not last long. Nobody expects a party's programs and promises to be taken as seriously as marriage vows once an election is over, but the ten programs on which the ten political groups got their deputies elected last month contain the seeds of the kind of squabble and disagreement that has produced chaos in Parliament over the long yesterdays, and can do the same tomorrow.

February 16

For Paris, this has been the longest, coldest stretch of septentrional weather known since the first winter of the Occupation, in 1940, and for Europe in general, the temperatures have been the lowest so far endured in this century. A mosiac of ice blocks covers the Seine, and the Tuileries gardens are dazzling with snow and sunshine. One of the dangerous freaks of the cold snap everywhere in France has been the blizzard drop in temperature at night and the warmish rise in the sun the next day, often with a swing of forty degrees in twelve hours. This sort of weather has made thaws, refreezes, burst water pipes, icy streets, motor accidents, broken legs, and pneumonia the regular news.

In Paris, braziers have been set up for customers on the café terraces along the Champs-Elysées and for the city's *clochards,* or tramps, on the Place Maubert, the ragpickers' mart. A fourth of the street traffic lights froze last week, but fortunately there was nearly no traffic, even the Place de la Concorde being almost bare of cars. Shipping is paralyzed on rivers and in ports. The Seine is blocked from Rouen to the sea; coal barges are stuck fast in the upcountry canals; even at Marseille one of the port basins is frozen shut, under brilliant sunshine and the fiercest mistral gale in years. The wind has been blowing at eighty miles an hour down the Rhône Valley, and the River Saône is frozen for twenty miles around Lyon. Over last weekend, the two coldest cities in France were Strasbourg, in the north, and Limoges, toward the south, both registering only eight degrees warmer than Moscow; yesterday we heard that Mulhouse had just beaten the Moscow temperature by six degrees, having dropped to ten degrees below zero. Another weather oddity is that it has been colder on the Riviera than here in the Ile-de-France. There is skiing on three feet of snow in the streets of St.-Tropez; snowplows have cleared the roads around Cannes; after its Battle of Flowers, Nice turned into a Christmas card overnight, with the worst snowfall since 1887. Along the rest of the Côte d'Azur, the fields of carnations are so badly damaged that prices have sextupled on the wholesale flower market in Nice. Around Pau, the famous crop of early artichokes for the Paris trade has suffered, and at Arcachon thousands of dozens of the succulent *claires* oysters have been literally frozen to death in their beds. Paris vegetable prices for last Sunday's dinner leaped sky-high. Potatoes jumped from sixteen to twenty-seven francs per kilo and soup leeks from a hundred and twelve to three hundred and fifty; lettuces soared from two hundred francs to four hundred and twenty, which meant that few families could afford a salad with the weekly roast; and truckloads of fish, poultry, and meat are stalled on icy roads all over France. Mounting food prices are an added irritant that French citizens and their government could well do without right now.

Socialist politicians here are generally regarded as too intellectual, too humanitarian dialectically, sometimes too truthful, and usually too inexperienced to be practical as government leaders in a crisis—almost the only time they are ever called on. It was humanitarian and intellectual of Socialist Premier Guy Mollet, in his radio speech in Algiers, to say to the eight million native Moslems, nine-tenths of them illiterate, "I recognize your material misery, the injustices you have known, your impression of being second-class citizens, your suffering in your dignity as men." And it was certainly not practical, no matter how true, to tell the French white settlers that some of them were "an egotistical small minority" bent on selfishly preserving its financial and political superiority over the natives. The mob violence with which the white French greeted Mollet's arrival in Algiers—pelting him with eggs, stones, rotten tomatoes, and dung—was a scandalous indignity that no French

Premier apparently has ever before experienced. But his failure to make a strong speech, for all France to hear, denouncing such dangerous anti-Republican mob pressure, and his even seeming to turn the other check by declaring that the "dolorous demonstrations" had informed him of the white settlers' attachment to France and their fear of being abandoned in appeasement of the natives—these things shocked many elements in Paris, including certain members of his own government. One Paris newspaper editorial jeeringly said that French politics is not a game for choirboys. François Mauriac contributed a denunciation in *L'Express* by quoting a scholarly line from Malherbe's "Ode to Louis XIII"—"Take thy thunderbolt and advance like a lion"—and then adding, in contemptuous slang, "Alas, M. Guy Mollet did not take his thunderbolt. He took it on the nose, in the shape of rotten tomatoes."

It was known in advance of Premier Mollet's presentation of his Algerian program to Parliament, which has just begun, that his main reform for the natives—an eight-to-one majority of the population—is a promise of equality in every way with the French whites, who up to the present, for many reasons and for all the privileges, have kept the natives as a depressed minority. What the Algerian native leaders reportedly want now is no longer equality but an independent Algerian nation, in which the French white settlers will be free to live as a minority.

The price has gone up during the last few months in all the French Arab lands, and this is the week of the triple climax in France's relations with her North African holdings. The restored Sultan of Morocco, Sidi Mohammed ben Youssef, in Paris on a state visit, has already clarified for President Coty his desire for a Sherifian Empire and eventual Moroccan independence. Habib Bourguiba, leader of the Tunisian Neo-Destour independence movement, has just held talks in Paris with the French government, and now the French High Commissioner for Tunisia has arrived here to report to his government that His Majesty the Bey desires new talks with the French, which means new Tunisian libertarian demands. Bloodshed, terrorism, and a state of small, awful civil war that involves the terrorist natives and the better part of the French Army continue in all three Moslem lands. This week is of the greatest importance in statesmanship for the future of Republican France and of an importance only slightly less for her democratic allies.

* * *

Everybody—even the French, in between times—always seems to forget the terrifying indigenous French capacity for violence in language and in physical scrimmages in the melodramas of ordinary political life. The new Poujade Party, in inspiring a revival of the cry "Down with Fascism!," has rekindled all over France those fiery partisanships of Left against Right inherited from the Revolution, and the Poujadists' filibusters, arguments, and losing struggle (so far) against the invalidation of some of their deputies, because of incorrect use of the election lists, have brought a new intensity into Parliament. In the quarrels that have developed, the parliamentary language exchanged between the Poujadists and the Communists—the moderate Center kept a more civil tongue in its head—has been appalling. Deputies have called each other Hitlerite, Muscovite, gangster, stool pigeon, bandit, whore, carrion, rotter, thug, and denunciator. This week's fist fight, which broke out between the extreme Right and the extreme Left, and then involved almost the whole Assembly in a struggle either to calm the others or to manhandle them, featured bloody noses and the throwing of chairs, with peace restored only by surprise, when an overexcited madman in the visitors' gallery fired four blank pistol shots. In the four weeks this new Parliament has been sitting, this fight was probably the most important session yet held.

February 29

This has recently been a stimulating theatre season for its extreme and interesting oddities—like Little Theatre experiments—probably staged because of the success, three years back, of Samuel Beckett's "En Attendant Godot," which set a style in the peculiar. Outstanding is "Le Personnage Combattant," by Jean Vauthier, the newest avant-garde playwright, at the Petit Marigny, an upstairs theatre and workshop. It is in two fifty-minute acts, with Jean-Louis Barrault alone on the stage and talking full tilt for forty minutes of each, in what is, for once, a tour de force of brilliant, egotistical, intellectual acting. He plays a disillusioned, unsuccessful writer, come back like a soiled homing pigeon to a cheap hotel room over a railroad station—the play's elaborate sound track of whistling locomotives furnishes a kind of mechanized Greek chorus—where, years ago, he wrote his only fine pages. These he reads aloud, and in starting to expand them, as if rewriting, audibly composes his sordid history, with its noble aspirations and admitted failure—his true short story as the would-be artist of words who cannot now write his life tragedy but can only talk it, in a lonely monologue. It is the most interesting play and acting in Paris.

The other popular unorthodox playwright is Eugène Ionesco, whose use of logic—first as a vehicle of folly, then as an agent of destruction—in his playlets has given him an admirable reputation (one is never quite sure just why) among intellectual theatregoers. His present playlet, a revival from three years ago, when few seem to have seen it, is at the upstairs Studio des Champs-Elysées and is called "Les Chaises." It shows an old married couple living in some remote tower by the water. The husband still has a message of salvation to give to a world that has never listened to him, and they invite a numerous imaginary company as listeners, for whom the chairs of the title are brought onstage. The invisible guests culminate with the Emperor himself, plus an orator who is supposed to read the great message but turns out to be deaf and dumb. As part of their triumphant farewell to the also deaf world, the old couple throw themselves into the water below. Well, there you are.

March 20

One of France's many worries in connection with the rising tide of the Arabs in North Africa has been Washington's silence as to which foot it is standing on—our historic anti-colonialism or our modern anti-Soviet policy that ties us to our allies and our North African military airfields. The decisive speech just made at this week's diplomatic-press luncheon by our Ambassador, Mr. C. Douglas Dillon, has been headlined in the Paris papers for his statement that France's search for liberal solutions to insure the continuance of the French presence in North Africa "has the wholehearted support of the United States government." In thanking him for "this remarkable intervention," *Le Monde* added, "One can imagine the resistance he must have overcome in Washington"—a reference to the rumor here that the Ambassador had to fight his way up through the State Department to the President himself for permission to make any positive declaration.

This week, the independence of Tunisia has been recognized by

France, as the independence of Morocco was two weeks ago. As French colonies, they have ironically and suddenly won where Algeria, regarded as an integral physical part of France located across the Mediterranean, is in the position of our South in the Civil War in rebelling to obtain the same independence; its *fellaghas* are fighting for secession, which is illegal. The Mollet government has now reached the peak of its effort in the dual consignments it is sending to Algeria, either to enforce peace or to negotiate it. It is transporting an enormous French army to fight the rebels, and it is repeating to the natives its promises of a better new life, of more government jobs, of agrarian reform, of higher wages, of more employment, and of fair elections. Had previous French governments offered the reforms two years ago, many French parents of sons being sent to kill or be killed in Algeria feel, the present government would not now be involved in this Moslem-Christian civil war.

March 27

The Communist daily, *L'Humanité,* is now running one of those circulation-boosting competitions (first prize is a three-week trip for two to the Soviet Union) based—unfortunately, at this moment in Communist history—on the old game of questions and answers, the competition being called *"Vrai ou Faux?"* Last week, *Figaro* pulled *L'Humanité's* leg by sarcastically listing questions of its own that *Huma* might well ask, such as "Stalin was wrong for twenty-five years. True or false?" and "Stalin was tyrannical, maniacal, and sanguinary. True or false?" It is certainly true that the February attack on the Stalin myth by Moscow's Twentieth Communist Party Congress, falling on the utterly unprepared and rigorously pro-Stalin French Communist Party, was the most destructive blow the Party here has suffered since its foundation in 1920—a blow from which it staggered into a recovery, at least officially, only last week. Another obvious truth is that the French Party delegates to the Congress—Maurice Thorez and Jacques Duclos—were not warned on their arrival in Moscow of the imminent anti-Stalin coup de grâce. Consequently, the wretched Thorez made a rousing pro-Stalin Congress speech so contrary to the coming new Soviet policy that if he had made such a *gaffe* (on a quite different line, of course) during Stalin's lifetime, the genial Georgian would doubtless have had him purged the next morning. With the temporary crackup of Party discipline here over the last month, many intellectual Communists are speaking with more bitter candor than ever before. They say that the French Party has for several years been ignored by Moscow in favor of the Italian Communist Party, which is more militant, more useful, and bigger than theirs and has a more intelligent, finer-brained leader in the highly educated, intellectual Palmiro Togliatti, who comes of a university family (his brother is Dean of the Sciences Department at Genoa University) than the French Party has in the former pastry cook Jacques Duclos, who is its dominating secretary. It is thought here that because Togliatti never once mentioned Stalin's name in his Congress speech, the Italian Party had surely been warned, as too valuable to Moscow, not to let its leader make a fool of himself there and so increase the difficulties of de-Stalinizing the Italian Comrades, who, having been accustomed over long centuries to saints, were inclined to especially easy credulity where the Georgian was concerned. In canonizing Stalin, the task of the French Party propagandists, dealing with the more cynical and more agnostic-minded French workmen over the last twenty-five years, had been harder. Yet they had succeeded in instilling in them a kind of glamorous, historical, deep devotion to Stalin as a remaker of the world—a devotion like the one their grandfathers felt for the myth of Napoleon. All that is gone now.

It took the erstwhile pro-Stalin French Communist Party leaders five weeks to make up their minds to eat crow. They swallowed it whole last Thursday. In *L'Humanité's* March 23rd front-page headline, the French Central Committee hailed "LE VINGTIEME CONGRES DU PARTI COMMUNISTE DE L'UNION SOVIETIQUE" as "UNE AIDE INESTIMABLE POUR NOTRE PARTI." On an inner page was given the French Party's resolution, arrived at by the chiefs during a twelve-hour closed meeting the day before. *Huma* printed the resolution in three scant columns (in Rome, chief Togliatti, possibly better prepared, had earlier given twenty-seven columns to his crow-eating in *L'Unità*), of which seven-eighths was given over entirely to "the superiority of Communism" and its amazing recent advances in technology and production, as reported at the Twentieth Congress. Only the last fraction of the resolution mentioned Stalin. There the French Party's humiliating capitulation was printed, in what sounded like fragmentary heretical *mea culpas*—"Stalin's erroneous theories . . . harmful consequences . . . violation of Leninist principles"—with the final solemn ending "Let us fight against the cult of personality." Not only had the French Party made a cult of Stalin but it had made an even stronger cult of its own Maurice Thorez as a *fils du peuple*—handsome, strong, a miner who was a miner's son, the ideal Communist proletarian who became the French Comrades' real idol and who led them deeper into the Stalin error at the Congress of Moscow.

This has been the first Paris sight of a Slav holy man's being ruled off the international Marxist saints' calendar—the first time people here have ever seen what would have been heresy in January become revealed doctrine in March to five million French workers, supposedly the best educated and most intelligent in Europe. It has been an extraordinarily unholy spectacle to watch.

Les Concerts du Domaine Musical, a series dedicated to contemporary and often brand-new music, have filled a real want here and lately have overfilled the Petit Théâtre Marigny, the Renaud-Barrault little-theatre workshop over the Théâtre Marigny, where listeners have been packing the hard benches and sitting on the floor with a lack of old-fashioned comfort suitable to such modern, angular compositions. These are the only concerts in Paris devoted mainly to twelve-tone music, which is considered a trifle démodé by other Europeans, since they have known it since Schoenberg's early days, but which has only recently struck French youth and, indeed, many French adults. The Domaine's final concert of the season, given last week, featured in its first half the 1928 symphony of the late Austrian composer Anton Webern and six of his songs (two of them written to poems by Rilke), and demonstrated once again his amazing endowments—his conciseness, his subtlety of musical language, and his originalities, which could actually be followed even by the musical layman. Webern's music and the more satisfying romantic works of Alban Berg—his "Wozzeck," given here by the Vienna State Opera four years ago, still ranks as a memorable event—make these two dodecaphonists seem to the French today more highly gifted composers than their master, Schoenberg. Born an aristocrat and a *von,* Herr Webern, as he called himself after he turned Communist, was *chef d'orchestre* of the Vienna Workers' Symphony Orchestra for many years. (He died just after the Second World War.) In accordance with his political decision, he was dutifully styled without his "*von*" on the Domaine program. The second half of the evening was given over to "Le Marteau sans Maître," which has poems by René Char as a text. It was a mixture of occasional soprano singing and continued instrumental comment by a flute, a guitar, a vibraphone, a xylophone, and little drums like flowerpots, the music having been recently written by Pierre Boulez, the Domaine's director and the leading French twelve-tone composer. His music is purely Gallic, purely stylistic, and purely brilliant in its invention and syncopation, and it was flawlessly executed, like everything by everybody that is played at these concerts. Other composers featured this month have been the important young Italian Luigi Nono, the German Hans Henze, and the irrepressible French experimenter Olivier Messiaen, in an opus composed of birds' songs transcribed into ancient musical modes to avoid commonplace majors and minors, with diverting scholarly results. Parisians ceaselessly criticize what they call a general letdown now in the performances, audiences, and programs of their once famous big classic orchestras. The Domaine concerts appeal exclusively to lovers of extremely modern music; no one else could bear to listen to them. Their dodecaphony is so popular and the Petit Marigny so small that the same program is always given two nights in succession. The concerts receive fine critical notices and discriminatingly enthusiastic applause. These make them doubly rare in the Paris concert field today.

May 20

The violence and horrors of the war in Algeria and the divisions of opinion on it here in Paris have been profoundly worrying the French. There has even been talk of reforms that have no bearing on anything right now but represent the general malaise, the restlessness of hopes, the onset of self-criticism. There is talk of copying the American Presidential form of government, so as to have a strong elected head, instead of unstable, helpless Premiers. There is some talk of a return to the highest office of General de Gaulle, sure sign of the temporary idealism that comes with deep national anxiety—the phantom of the incorruptible strong man, who would now consent to govern only if he could start from strength, not from the weakness of the present form of government. The drab Champs-Elysées military parade that recently celebrated the eleventh anniversary of the ending of the war gave a picture of France's deprivations these last months in North Africa. There were no colonial troops in the parade. General de Lattre de Tassigny's famous Atlas Mountain *goums,* in turbans and on galloping stallions, have been enrolled in the Sultan's army in Morocco, now an independent Sherifian state, where natives have been burning each other alive. There were no regiments in red fezzes from Tunisia, also newly "independent in interdependence," where massacres have been continuing and where Premier Habib Bourguiba has just demanded the right to send Tunisian diplomats to Washington and London and to establish a Tunisian Army. There were no white-capped, bearded Foreign Legionnaires, now fighting in Algeria, where all forms of horrible death are part of the war. Men have been mutilated, children have been stabbed, throats have been cut, farmhouses and farm crops, and their owners, have been fired by torches and gasoline, with natives fighting the whites, whites fighting the natives, and some natives fighting other natives. Because, after hundreds of years, history has come to a sudden head on the south side of the Mediterranean, Premier Mollet, Socialist and pacifist by training, has been forced to do in Algeria what previous, reactionary French governments neither dared nor wanted to do. He has been making a war of pacification with one hand and, with the other, offering drastic reforms to eight and a half million roused Moslem natives, all the while also trying to preserve the existence there of the one million colonial whites—French, Italian, and numerous Maltese—for whom he and France are responsible today. Many French here in France, recalling Indo-China, think that nobody can prophesy how this racial war between two peoples at such different stages of civilization will turn out. Millions of the French in France passionately, patriotically believe in the righteousness of the war and in its victorious outcome. Other millions believe that history in the past ten years finished the colonial cycle of all Western liberal peoples (thus excluding the Russians). And still other millions are Communists, to whom the Algerian war is *une sale guerre*. Though the Communist deputies in Parliament supported Mollet's demand for special powers to fight it, they are now agitating for negotiations to stop it with a cease-fire. The forceful Robert Lacoste, Resident Minister in Algeria, has just stated that "though I promise no miracles, I have a reasonable hope that by the end of summer law and order will reign in Algeria."

What discourages so many French is that law and order have not yet started to reign in independent Morocco and Tunisia.

May 29

Eleven Frenchmen and one Frenchwoman had their throats slit last Friday by Algerian rebels a few kilometres south of Biskra, in the Sahara, the desert once featured in "The Garden of Allah," that ancient best-seller that touched on tender relations between natives and whites. Earlier in the week, outside Philippeville, two Moslem families—among them seven women and seven children, one three months old—suspected of loyalty to the French also had their throats cut by rebels, who, furthermore, decapitated their victims' chickens. Most of the Moslem students at the University of Algiers have abandoned their classes, well before their June exams, in response to a rebel slogan declaring, "Examinations make no sense today! Join the Maquis against the French!" Ten days ago, in the gorges near Palestro, seventeen out of twenty-two green French soldiers, in their third week of war, were ambushed and massacred by the rebels they were hunting. After certain Paris newspapers printed not only the names of the boys—eleven came from small towns near Paris—but the horrifying particulars of the mutilations and tortures that killed them, a wartime censorship was set up, forbidding the publication of "morbid details." Premier Guy Mollet is awaiting the results of a three-day Parliamentary debate about the Algerian situation. Grave as that situation is, he is not expected to fall, for though his Rightist political enemies think the war is going badly, no opposition party wants to be nationally responsible, as the Socialists are now, for trying to make it go better—even if any opposition party were sure it knew how. Finance Minister Paul Ramadier has just warned the public that the Algerian campaign will necessitate immediate new temporary taxes, the last adjective being the only part of the statement that sounds optimistic.

This is a cruel, fanatical, démodé Arab holy war, an intimate, hand-to-hand war of native knives and barbaric tortures, in which French helicopters supply the outstanding modern touch, a war of *petits paquets,* with the communiqués mentioning such small parcels of men as to sound ludicrous—except that they are descriptive of the endless, frittering kind of hide-and-seek, hill-and-desert war it really is. One day's communiqué last week featured engagements at Lourmel (rebel band surrounded, fourteen prisoners), Gambetta (three rebels killed), El-Kseur (nineteen rebels killed), and Aurès-Nemencha (skirmish with rebel bands, eleven wounded, two French military killed)—a total of twenty-two dead Algerians and two dead Frenchmen, at a cost of millions of francs, for that day.

The resignation of Pierre Mendès-France from Mollet's Cabinet—where, unfortunately, he had been doing nothing anyhow—brought to light again his "seven significant measures," which he announced in April, and which he still thinks can alone "save the French presence in Algeria." Practically all the French colonials there, and many Frenchmen here, think that if these measures were applied, there would be nearly no Algeria worth saving, as far as French interests go, so why shed blood for such remnants? (An increased war effort is part of the Mendès-France program.) As an economic expert, Mendès proposed the expropriation of all sizable agricultural properties, in order to turn them into small family holdings, and a fundamental reform of agricultural credit that would "extract it from the selfish hands of big owners, who have always profited at the expense of little producers." He wants raised wages, recognition of Moslem labor unions, freedom of opinion for native Arabic newspapers, and the removal of those anti-native French functionaries who from the start have kept everything in their own white hands and out of Arab reach—all this to "promote native confidence and hope in France, without which, sooner or later, we French will be evicted from Algeria and from all of North Africa." Most scandalized French politicians look upon this as a program so bold that only the daring Mendès would risk trying it out, if he should become Premier—an event they are determined not to let happen to France.

Apropos of Mendès' resignation, *Le Monde* dryly remarked that he "is not an accommodating man or at his best except as first fiddle." For the nine years between 1945 and 1954, he played no solo part in Parliament except as a constant, ruthless—and by the outside world unknown—critic of all the French governments. Upon his sudden emergence, two years ago, as France's strong man, he roused the greatest hopes and the greatest devotion—and also the greatest personal hatred on the part of many—of any new French leader of modern times. Now both the former strong men who rose to save France after the war are off the stage—de Gaulle and Mendès-France, two most complex characters, and two incalculable losses for France in her present troubled hour.

July 5

The new novel, if it can be so called, by Albert Camus—"La Chute" ("The Fall")—is fascinating, strange, and brief. It consists of two hundred and sixty-nine pages of monologue by the only character in it; is the single consequential work of fiction so far this year by any of the French intellectuals, new or old; has already achieved magnetic popularity among general readers; and ranks as an imaginative creation on the same high and unhappy level of human inquiry as "La Peste" and "L'Etranger." Called a *récit,* it is a one-voiced confessional report on the character of a former Paris lawyer named Jean-Baptiste Clamence, who, in a moral shipwreck, has come to land in Amsterdam, in a sailors' bar. A chance French visitor "of the same race"—that is, also a lawyer—runs into him there and hears the *récit.* Clamence's story contains one sole anecdote: Having always defended only noble causes in the Paris courts, where he ranked as a moral dandy in the legal profession, he is promenading alone by the Seine one night when he hears the sound of a body striking the water and then a woman's voice crying for help—a repentant suicide, whom he indifferently lets drown. From then on, choosing his helpless clientele in this low seafarers' bar, he practices his new combined professions—those of a godless priest and father confessor, a judge, and a prosecutor, who prosecutes himself ceaselessly for his crime of omission, wrings confessions from the sailors of their crimes, and judges everybody on the theory that a judge must be guilty himself to understand and assay guilt. Obviously, what Clamence has arrived at is his own version of the fall of man, of what he considers to be God (in whom he does not believe), and of man's destiny, which cannot be pure and which makes both the Deity and mankind eternally guilty. Camus's fictional monologue is thus really a philosophical dialogue with life itself—different from Malraux's heroic inquiry, yet also based on proof by noble action; different from Sartre's Existentialist answer, yet also based on man's helpless absurdity. "When we are all guilty," Clamence says, finally, with mad and subtle lucidity, "that will be democracy."

July 11

The second volume of General Charles de Gaulle's "Mémoires de Guerre," covering the years 1942–44, has now been published. It is selling less well than the first volume, which became a best-seller largely through public curiosity, but its high literary style, matching de Gaulle's exalted patriotism—and, indeed, his physical stature, as if the latter permitted him a superior, lonely view—allows this second record to become a classic, indifferent to popularity. His historical position was strange and unique, like his Gothic nature and vision, and what he records is as individual as his experiences. "As for human relations," he writes poignantly, "my lot has been that of solitude." The memoirs are, signally, his story of his and France's history. They lack the universality and humanity that could perhaps be characterized as the genial reek of cigars—part of the smoke of the democratic battle—which hangs over Sir Winston Churchill's books on the same war. De Gaulle's aim as leader of the Resistance is summed up for himself and for his comrades of that time in the sentence "We bring back to France independence, the Empire, and the sword." A few days ago, when asked by a French journalist to make a statement on the present situation in the French Empire, de Gaulle said merely that he had made so many statements over recent years that a new one would add nothing and have no effect. This is tragically true, and the sadder for being stated by him.

September 5

The gravity of the Suez Canal situation was not lightened here by the Compagnie Universelle du Canal Maritime de Suez's making a fool of itself. Ten days ago, the company's main office, which is in Paris, mailed a check for a hundred thousand francs (about two hundred and eighty dollars), accompanied by a mealymouthed but indiscreet letter, to the Progressiste morning newspaper *La Libération,* which promptly printed photostats of both on its front page and ran a biting, jocular editorial that should have made the faces of the French stockholders in the Canal Company turn scarlet. The letter, which was written in the first person and bore the rubber-stamped signature of the *secrétaire général* of the Paris office, said, in effect, that ever since the Egyptian government nationalized the Canal, the company had had to issue press communiqués and hold press conferences aimed at "reëstablishing the truth and enlightening public opinion about the company's point of view" and went on, "I have not failed to realize that in some cases the publication of this information may have entailed expenses that I consider it legitimate for us to bear in part. In consequence, please find enclosed a check for one hundred thousand francs, as a contribution toward your August expenses. It is understood that our part in meeting them can be continued if the eventuality arises." The letter closed with the optimistic phrase "Hoping you will agree with us," as if it were a perfectly ordinary office communication, and not dynamite.

Paris was scandalized and disgusted. For the first time, the Canal Company lost face with some of its bourgeois and conservative admirers. The next morning, *La Libération* ran an editorial that righteously fulminated against this "attempt to bribe the press" and demanded to know why the rest of the Paris papers—which, the editorial writer said, must also have received checks—had remained mum, and it also printed on its front page what the editor called "the Suez Canal Company's confession," which did not amount to much. In it, the *secrétaire général* simply said that the whole thing was a regrettable mistake—that the company had thought it had to pay for news in newspapers, just as it did for its financial announcements on the stock-market pages, because it was "inexperienced in dealing with the press." This was a breathtaking admission of incompetence at a moment when the company's fate and that of the Suez Canal, which it has administered for nearly a hundred years, were making front-page news all over the world.

The most unexpected feature of the Suez crisis has been the continuing lack of nuance in the way the French react to it. Uncharacteristically, all the French seem to have managed a real union, for once—all militant, aroused, convinced, and all behind the government in its taking of what it openly calls its "extremely firm stand" of being prepared for military action, if necessary. All the French, that is, except that pro-Communist quarter of them that has been cheering for Colonel Nasser. Aside from the small Communist press, all the papers you read, which ordinarily feature dissenting convictions, and all the French people you see, whom you formerly listened to for their differing opinions, have been printing or saying essentially the same thing, in an impressive sincere chorus of extreme national sentiment. There has been no moderate note, no voice of opposition. For the past three weeks, those members of the American colony here who have wanted to know at least what opposition sounds like—and who have also realized that the Suez struggle concerns England no less than it concerns France—have been reading the London *Economist* and the Manchester *Guardian*. On Monday of this week, as the Cairo Conference opened, these publications were finally given major French recognition, in a *Figaro* editorial, as "two of the principal organs of the British press, the most notable weekly and the most respected daily of Great Britain," but both of them, unfortunately, "partisans of resignation"—that is to say, against immediate shooting and in favor of trying to end with nothing louder than words the sudden Mediterranean dilemma that history has appallingly brought to pass.

Paris is united principally by what it calls "the Munich complex." It is haunted by its twenty-year-old memories of the Führer and the nationalized Rhineland, which recently led to Foreign Minister Pineau's anguished inquiry "Is it not our duty to stop Nasser immediately?" Nasser is considered a Fascist here, and is automatically treated as a psychotic, rather than a politician. His Fascistic paranoia seemed proved to the French when, like the Führer after the Rhineland annexation, he turned toward Russia as his next move. It is to prevent his Pan-Arabianism from igniting the world the way Hitler's Pan-Germanism did and starting a great war that France appears willing to risk a small war. That usually reserved evening journal *Le Monde* declared in its Monday-night editorial, "The idea is firmly held here that stepping back now could constitute a more fearful menace for peace [than forward action]. To this is added the belief that the risk of the conflict's spreading would be graver in a few months, when the United States (its Presidential election over with) would have fewer motives for staying out."

While the talking at the Cairo Conference was getting started, the French began to figure out what they would have to face if, owing to the Suez crisis, gasoline should be rationed, or even cease to exist altogether—imaginatively suffering in advance that nightmare in which one tries to run but cannot move. France has a three-month supply of gasoline on hand, two-thirds of it in reserve tanks. The National Printing Press has denied that it has received a government order to print gas-ration coupons, adding that even if it had been so instructed, it has no watermarked official paper to print them on. The only good gasoline news is the recent announcement by the Compagnie de Recherches et d'Exploitation Pétrolières au Sahara that oil has been found in the Sahara Desert, at Edjellé and Tiguentourine. As late as a few years ago, when France still had a free hand in that desert waste, nothing much was heard of what is now called "the Sahara's hidden wealth of minerals and oils." Even today nobody knows the exact boundary line between Algeria and Morocco in the lower Sahara region, for the desert sands, roamed over only by wild tribes, were never considered important enough to survey.

September 18

Summer has finally come to Paris, in the early autumn. There are lovely cool nights that seem to be in communion with the calendar, and then sunny, warm July days, with the trees already carrying the colored foliage of October—a mixture of weather and seasons that passes everybody's comprehension and most Parisians' recollections but that serves well indeed for pleasure, for a feeling of holiday here at home at last. The farmers, who did not cut their wheat when they should have normally, in mid-July, are now making their harvest in mid-September, mowing and threshing and bringing in a fine retarded, rich, ripe, dry crop. No one has ever seen the seasons so turned about and around, but never has a St. Martin's summer been more welcome than this, which should by rights come in November.

The shoe is on the other foot now. When the dangerous Suez Canal crisis began, Paris loudly complained that Washington was manifesting too little concern over the vital matter, and today Washington is showing so much concern that the French wish it would let up somewhat. A plaintive editorial in *Combat* this week opened by saying, "As on most of the days that God grants us, M. Foster Dulles had something to say yesterday."

October 9

After a fifty-nine-day vacation, the French Parliament has opened again, with France on exactly the same two spots it was on when Parliament adjourned in August—the unsettled, grave Suez crisis and the worrying, unfinished Algerian war. As the influential evening paper *Le Monde* pointed out, "The Parliamentary reopening took place in a singular atmosphere. There was a contradiction between the public declarations and the private comments, and a contradiction, moreover, between the serenity and optimism of some and the reality of certain events these last weeks." The serene optimists are those who agree with Robert Lacoste, Resident Minister of Algeria, who stated on opening day that the twenty-three-month-old war in Algeria is surely being won, that the French Army only hit its stride in September, that October might be the bloodiest terrorist month of all, but that in November things should look brighter. One of the recent troubling realities was the latest Suez divagation, uttered at last Tuesday's press conference, by Secretary of State Dulles, who, *Le Monde* sharply said, "overstepped himself by airing the differences between the United States and its allies," and who also struck irritated sparks in Paris and London governmental, diplomatic, and editorial circles by his reference to colonialism. Secretary Dulles's further, semi-idealistic press remark that good might even come of the Suez difficulties if they stimulated European federation brought a grim smile from French politicians. It is true that lately there has been revived talk on the Continent about a United Europe, but the feeling of union unfortunately seems to be founded largely on a common dislike of Mr. Dulles.

The major Lacoste news was his announcement that two hundred thousand acres of private Algerian property—farm lands belonging to the Compagnie Algérienne and to a Swiss company—have just been handed over to the Algerian agrarian-reform bureau as the first gesture toward dividing up certain nineteenth-century domains larger than twenty-five hundred acres into small farm parcels for land-poor natives, a French pacification effort. Another top announcement on Parliament's opening day was that the entire issue of government bonds to fight the Algerian war has been subscribed, and that the sale is now closed. A hundred and fifty billion francs had to be raised if income taxes were not to be boosted; actually, three hundred and twenty billion francs' worth of bonds were sold, though the issue was on the market only three and a half weeks. Bankers say that this was the quickest—and thus the most successful—national war loan ever undertaken, even including the one with which the French spiritedly paid their War of 1870 ransom to the disappointed Bismarck before it was due. Bankers have also pointed out that this recent loan was the most tempting investment and convenience any French government has ever dreamed up. (Its lures actually frightened off some people, who suspected a tempting trap.) The loan bonds pay five-per-cent interest, their redemption price is tied to an index of selected leading stocks on the Bourse, to prevent their devaluation, and the income from them is tax free for five years. After Nasser's nationalization of the Suez Canal, there was a heavy flight of French capital to Switzerland and New York, in an effort to recoup losses on the freezing of the Suez shares—a disaster for investors here, since they paid eight per cent, now only a golden memory. Since mid-September, many really rich French have sold their American shares and put the money in the French war loan—an ideal mixture of patriotism and attractive investment. This has certainly been a surprising Socialist government in most people's eyes. Socialists are supposed to be fuzzy about finance, and are theoretically pacifists. Socialist Premier Guy Mollet has just raised the fastest state loan in modern French history to help pay for a still unfinished rebel war that the preceding conservative governments did not have the courage to tackle.

Though Jean Dutourd's "Les Taxis de la Marne" has been out only a month, it is the book that is the most talked of, for and against, in Paris now. Dutourd, in its final lines, says that he gave it its title to recall the engagement with history that was "the most glorious—and least miraculous—of the twentieth century." This provocative essay of close to three hundred pages is a critique of France today and yesterday—an angry inquiry into what happened to the fibre of the nation and that of its Frenchmen. In part, it is a comparison between the France of September, 1914, which had the spunk and imagination to taxi its reinforcements of *poilus* to the Marne battlefield to save the country's life, and the France of June, 1940, when "the Generals were stupid, the soldiers did not want to die," and France was lost. Himself aged twenty in 1940, Dutourd says that, like most young men of his generation, he was a soldier without patriotism and without fight in him—or even the need for it—during his military career, which lasted fourteen days exactly, while France took time to fall. "It was not the hour for courage. France had forgotten the word," he says. His callow indifference to victory and to *La Patrie* he blames on the people who made the climate in which his generation grew up: the pacifist generals, the anti-war bourgeois men of the Left, the government men who wanted anything they could get except responsibility, the mediocre men like the lachrymose President of France, Albert Lebrun—"the man of destiny of the century." With heavy irony, he says that they robbed him of possibly being a hero and of maybe being a youthful captain with a Legion of Honor rosette for his bravery—"Ah, those bandits!"

François Mauriac, in his column "Bloc-Notes," in *L'Express*—the most carefully followed column of Paris—has said that Dutourd's is "an important book, perhaps less for what it clearly tells us than for what it represents," adding, "It is an outcry, at least. Someone has raised his voice." Dutourd says that he confined himself to inscribing "a phenomenon of which I myself was the source"—that is, his slow development into the patriot he is today, "unrecognizable in my own eyes, incarnating in the extreme everything I had despised eighteen years ago." The value of his confession, his critique, his anger, and his grief is that his book reads like a man talking out loud to himself; it is an account made up of little remembered things from his history or France's, from his mind, from Voltaire's mind, from the mind of his sergeant in the war—a book that is confidential and emotional, an analysis of remorse, memories, hope, and late-born love of his country. Searching to find out why patriotism was early killed in his generation and why it is still being murdered today, he lists what seem to him to be the French people's favorite weaknesses—their narcissism, their lack of faith, their egotism, their frivolous anarchy, their individualism that constantly drives them to nonconformism, their false myth of their "glorious defeat" (since defeat without an adjective would be unbearable to their pride), and their concurrent rival myth of French panache. His frame of reference is mostly the week that he and four soldier comrades, led by their sergeant, spent tramping around Brittany on a comfortable promenade of inglorious defeat after France fell. It was then, listening to a middle-aged barmaid, that he first heard the name of General de Gaulle—to Dutourd today the only man of honor, who brought back from England the little that was left of honor for France, and has taken the even smaller quantity remaining today into retreat with him, almost as if it were his personal property. Literary and intellectual French circles have been able to disdain Dutourd's book as too popular. Because he brings up no political problems of today, other circles have accused him of trading on the nonpolitical tricolor. But nobody accuses him of not having written something that the French of all classes, and even American residents here, can understand—a handbook of patriotism. This is why it has been the most talked-of book of the month.

November 4

October, with its violent historical surprises, has partly altered the Western world. Even those things that may turn out to be only temporary—the successful liberalization won by the Warsaw Communists; the wild, inspiring revolt for freedom in Budapest, now drowned in its own blood; the crumbling of Cairo's troops before the soldiers of Jerusalem; and the Paris and London unilateral move for independence from Washington—already have their own form of permanence in the new memories of European men and European governments, being a private file of parts of a new pattern for this hemisphere. It is widely considered here that October brought rebellions, of obviously different types, against the two rival power leaders of the West—the Soviet Union and the United States—now both gravely weakened in their influence. Neither of them, it is said, had been giving enough leeway to its captive or allied associates, who have now made a bolt for freedom in order to manage their steadily worsening affairs more profitably—they hope—and, most important, according to their own national necessities and lights. These last months can be called, in all historical gravity, the period of the worms that turned—Gomulka, Nagy (now in prison), the Sultan of Morocco, Premier Bourguiba of Tunis, Ben-Gurion of Israel, the amateur, intransigent dictator Nasser, Premier Guy Mollet, Sir Anthony Eden, and, finally, Sir Winston Churchill, in his intimate backing of the British Government's necessity for doing what it has done.

A vital cross-Channel footnote to this week's events is contained in the fundamental difference between the British Parliament, with a

Government acting for Her Majesty, and the French Chamber, representing utter republicanism with nearly a dozen negotiable parties, in their debates on the ultimatum to Egypt and the prospective invasion. Whereas the verbal, intellectual conflict between Opposition and Government in the British Parliament resounded worldwide, the Paris Chamber reacted with a laconic, if favorable, monotony that was its only distinction. On Tuesday night, its visitors' galleries were packed and tense and its hemicycle was richly filled with former Premiers and other officials, all come to hear what the French Assembly's reaction might be to the big event—Mollet's earlier declaration of the ultimatum to Egypt. A hundred and forty-nine Communist deputies voted against Mollet, as well as twenty-eight Poujadists—Poujade himself (who runs his party from outside) having phoned in for them "not to vote for a war to help the Queen of England," which offered the only rewarding moment of the evening. As for the Opposition, Mendès-France, who leads it, abstained from speaking, abstained from voting, and even abstained from applauding Mollet—"to mark his anxiety and disapproval," one newspaper sententiously remarked. There was literally no dissent raised against this critical, this possibly dangerous, government decision ("a throw of the dice," *Le Monde* called it, accepting the gamble and underlining its necessity) except the purely formal "speech of response," which Chamber law imposes after a government declaration. This response speech was astutely allotted to a Communist deputy, as a lightning rod on whom any Parliamentary distemper would naturally strike. His untimely Muscovite references to "colonial tyranny" naturally aroused heckling deputies, and shouts of "And Budapest?" and "Tell that to Stalin!" were the evening's only protests. Shortly afterward, Mollet, in asking for a vote of approval, said, "I have confidence in the patriotism and wisdom of the Assembly"—which showed it by voting three hundred and sixty-eight in his favor—and the strangely muted great Parliamentary debate was finished.

The two important explanations for its muteness are, first, that ever since Nasser nationalized the Canal, the French have been united as never before in wishing both to punish him and to save themselves from an Arab Hitler, as they did not save themselves from the Austrian original in 1938, and, second, that with the Socialists as the government, though unsocialistically acting according to their own ideology, there remains not one French party to speak as grouped oppositional liberals—a viewpoint that was lost the other night in a chasm of silence, without even an echo from the tempestuous events in England. However, the result was as the French truly wished it—an ultimatum not only to Egypt but to their own loss of pride, leadership, and energy, and to their discouragement and, indeed, bitter disillusion with Washington, Dulles, and the United Nations.

Owing to the air-raid arrest of the five leaders of the Algerian National Front, now imprisoned in Paris, the Sultan of Morocco has declared that the humiliation to his sovereignty can be appeased only by an explanation from his French equal, President René Coty, who, since he disapproved heartily of the air piracy, could say little. Premier Bourguiba has declared that he and the Bey of Tunis have now taken a public position against French military action in Egypt. "The hatred of the Moroccans and Tunisians will be redoubled against us, because of Algeria," François Mauriac, the great Arab sympathizer, justly wrote in his column "Bloc-Notes," in *L'Express.*

But more consequential this weekend than the fight against Nasser, or than sympathy for ill-used Israel, or than anxiety about their own troubling North African problems, has been the French emotion over the anguish in Budapest. There are no newspapers here on Sunday—nothing but the radio. In its noon announcement today, the Budapest rebel station broke off, saying, "We are now going off the air. *Vive l'Europe! Vive la Hongrie!* We are dying for Hungary." Then silence.

November 8

The French and British so-called armed police action in Egypt lasted only from six-thirty Wednesday evening until midnight Tuesday—the shortest, gravest, and most unsettled (and still the most unsettling) military campaign in the immediate history of our time. The French people and the French government, intrepidly united six days before by the Suez action, were indescribably relieved—like the world at large—by the cease-fire. During the six-day Egyptian war, as it was called, the French, who, because their Army is largely engaged in Algeria, were able to supply only a third of the Anglo-French armed forces, were disproportionately satisfied with the communiqué by their Secretary of State for Air that between Thursday and Sunday their fliers had "carried out more than five hundred assault and reconnaissance missions, without a single loss." And, after the official blackout of news that began in the first minutes of Wednesday morning, they were semi-satisfied with Premier Guy Mollet's announcement that the Anglo-French forces were installed "temporarily in key positions on the canal at Port Said, Ismailia, and Suez." Only belatedly have they discovered that the troops are nowhere near that far but are, in truth, only about thirty miles down the Canal, now blocked with the ships and dredges that Colonel Nasser triumphantly sank—his big naval victory. Mollet has just gravely declared, in totting up the final six-day account, that "there has been more gain than loss" in the explosive, startling, and short Egyptian campaign, which turned out to be too short for success by a few days—maybe by a few hours—while the world held its breath. "France and Great Britain had swallowed an impressive number of slaps in the face and kicks before taking steps in Egypt," *Le Monde* sombrely remarked. As for the United Nations' moral censure, the paper said, with cruel realism, "The nations sitting in judgment there apply to others principles they would not follow if they were in a similar situation. The Soviet Union condemns France and Great Britain for what they have done in the Near East, but it has done worse in Hungary. The United States has clean hands and a pure heart in both cases, because the United States is not involved, but has it forgotten what it did in Guatemala?"

As seen here, the single and tremendous result of the Egyptian campaign—one that can possibly have an immediate effect on the world and history—is that the United Nations will now have its own international armed force, its own mixed army, to fight against fighting, whose first contingents are expected to be assembled this very weekend with haste and realism, stimulated first by fear and now by hope.

The French flag on the Paris Hôtel de Ville has been flying at half mast to honor the Hungarian dead. In the sudden, unexpected glut of bloody history, with two unofficial wars on hand, one of them embroiling the French, the reaction of Parisians has never swerved. Regardless of what their Army was doing in Egypt, their emotions were fixed on Hungary's fight for liberty. All available news of it overflowed the newspapers and filled the air on radios and on television, where hanged patriots were seen, pendent from a Budapest bridge. The latest boulevard newsreel shows parts of the shattered city itself; it looks like the ruins of Cologne or Berlin at the end of the Second World War.

As one of the odd coincidences that have added to the tension here, a Budapest circus (as well as a dainty circus from Peiping) was playing in Paris when the revolution broke. At a gala charity evening, it raised two million francs for the Hungarian Red Cross while the lions roared, some of the acrobats visibly wept in mid-air, and the clowns vainly tried to be gay. The Municipal Council of Paris has given ten million francs to the French Red Cross for transfer, which includes four million it had intended to spend on a coming visit here of Russian officials, now cancelled. At the reception held Wednesday in the Russian Embassy, on the Rue de Grenelle, to commemorate the thirty-ninth anniversary of the 1917 October Revolution—the old-style Czarist calendar making the difference in the dates—the French diplomatic corps, French officials, and all but definitely pro-Communist French guests failed to appear.

The French feel that, in crushing Hungary, Soviet Russia has destroyed not only Budapest but its own tentative place in civilized European history. "The Soviets' brutal, bloody repression of Hungary"—the customary phrase in shocked press clippings from around the world—also shocked certain Communists here, as important a happening as the first leak in a dike. A few talented literary Party members, like Louis Aragon and his novelist wife, Elsa Triolet, "though deeply divided in interpreting recent Hungarian events," as they declared in a manifesto, were united enough to ask Hungary's new Muscovite Premier Kadar not to kill or imprison the rebellious patriot Hungarian writers and intellectuals—"carriers of a part of human culture." Some former Communist-inclined *progressistes,* including Simone de Beauvoir and Jean-Paul Sartre, protested to the Soviet government against using "cannons and tanks to break the Hungarian people's revolt," adding that nobody could join their protest who had been "silent when the United States snuffed out the Guatemalans' liberty in blood, or who had applauded the *coup de Suez.*" However, the major single revolt was by Sartre himself, in a ten-thousand-word analysis in this week's *Express* of why, after Budapest, he is no longer pro-Communist—an amazingly calm, quiet, clear document. Among other things, he says that the Budapest crime lay not only in the tanks but in the fact that, from the Soviet viewpoint, they were necessary "after twelve years of terror and imbecility." He says that what the Hungarians have taught with their blood is that Marxist Socialism, as "merchandise exported by the U.S.S.R.," is a failure; that he thinks the de-Stalinization plan was sincere and courageous, though Khrushchev's anti-Stalin speech was in itself folly; that friendship can no longer be felt for the Soviet leaders, who now predominantly inspire horror; but that the path of mankind to the Left, although uncomfortable to tread, and maybe impractical, is the only way. He ends by saying that if the French Party calls him a jackal and a hyena (as it already has), this is "a matter of total indifference to me, considering what they called the events in Budapest," which was "a Fascist uprising." What he has written is a great modern corrective pamphlet.

December 7

After two years of remarkable prosperity, this season should be bringing a rich Christmas here. Instead, the French will be fortunate if they receive a few tankers of high-priced American oil. It is already urgently needed to keep men and industry at work in this part of the world until the Suez Canal flows freely again, perhaps in the spring. Politically, this oil has already cost dear in terms of weakening the Western alliance. The French feel that President Eisenhower has been doggedly using oil as if it were a kind of club held over his Franco-British allies—either they withdraw their troops from Egypt as a sign of wrongdoing or they will be punished by having no oil to keep business alive—and that this was only one of the many moralistic American attitudes during the Suez confusion that have jeopardized the Western ties. The possible loss of America's friendship was something that a multitude of French and British citizens—at the height of the recent anti-American bitterness—wished their countries could afford. For the past fortnight, everybody here except the Mollet government has known that in this Allied sector of the oil struggle the Franco-British pair would be beaten—that the moral General Ike and the amoral Colonel Nasser would win. Because of these two men's separate geographic strangleholds—as a result of which no oil from the old friend would come across the Atlantic, and none from Arab enemies could pass through the clogged Suez Canal—the troops are being hurried out of Port Said this week. Perhaps there will be some American oil for Western Europe and the British Isles by at least New Year's Day.

Thus comes to an end that earnest flight from reality which was the brief Franco-British invasion of Egypt, a lunatic, destructive slice of history hopefully set going by two sensible, highly patriotic men—Sir Anthony Eden, political pupil of old Sir Winston Churchill himself, and the humbler, Socialist Premier Guy Mollet, anti-Socialistically waging his second African punitive war. As the French clearly see now, the consequences have been catastrophic. Not only has Britain been hard hit financially but they themselves must face up to war costs, probably having to foot the bill for cleaning up Nasser's damage to the Canal, and losing the dollar credits they need for Texas oil, so much dearer than the Mohammedan pipeline variety. In addition to all this are the facts that the Franco-British soldiers who marched victoriously into Egypt and are now marching out again in defeat had only a toehold on the ruined Canal; that the oil is stopped; that every Arab has been turned into an enemy; that Russia is now firmly in the Middle East; and that the United States and the United Nations are so angry that, with brisk Soviet help, they have sanctified Nasser in the eyes of the world as the injured party after (in the eyes of the French and British, at least) he burgled their Suez Canal Company, which ably served the whole globe (Israel excepted). The cynical French being more realistic than the English, no bones were ever made here about what the French hoped was the Egyptian campaign's real aim—to throw out the Fascistic Nasser and imperiously seize the Canal for the democracies.

The French, having supposed that Sir Anthony Eden would disappear from power as a result of Suez, were agreeably astonished by the Conservative Party's victory last night in the House of Commons, but there has been no serious notion here that Premier Guy Mollet, though involved in the same Egyptian enterprise, would not continue as head of the French government. Incredibly enough, not until the week after next will there be a full-dress Parliamentary debate on his foreign policy, Suez included. Though England and France were militarily united in action, what happened in their countries and Parliaments from the beginning of the affair to the end of its aftermath yesterday was as different as day and night. It must be realized that during the six-day Egyptian war, and ever since, the biggest news on the Suez affair printed in the French papers has always been the British battle of public opinion over it, splitting the House of Commons and the British press and people—that already historic fight against Eden's pragmatic Conservatives and their war, waged, according to various shadings of conscience, by the British Labour Party, which is precisely the opposite number of the French Socialist Party, led by Eden's war partner. In addition to the irony of an alliance between France's supposedly Left government and the aristocratic English party, what must also be recognized is that the French, as a people and as a Parliament (Communists and Poujadists naturally excepted), were united behind Mollet's Egyptian campaign with a unanimity literally never seen before in modern France—as behind a kind of energetic Crusader's dream of righteousness in arms and of pursuing the treacherous infidel. The only violent public opinion aroused here that was comparable to the reaction in England was the unified, bitter conviction that America and the United Nations—with Russia—stopped the crusade in disapproval of its imminent victory, since that victory would have proved the garrulous inefficacy of the U.N. in bringing Nasser to terms, the unreliability of Mr. Dulles's multiple views, and Washington's lack of leadership in the Near East. Dulles's arrival here this weekend for consultation is awaited with grim interest.

As for the French foreign-affairs debate, those deputies who abstained from opposing Mollet's Suez project will, of course, belatedly voice their criticisms, and he will certainly be once more berated for his conduct of the ever-present Algerian war. To many of the French, that war and its possible localized military victory seem almost meaningless now, with the whole Arab world, including Morocco and Tunisia, fervently united against France since the Suez campaign. If Mollet remains in power, it will be because no other French political party is yet willing to relieve him of his accumulated burdens.

1957

June 18

Anyone who has been absent from Paris for six months and has just returned finds France in practically the same situations she was in last December (except that, unfortunately, they seem worsened by solidification or repetition)—finds the same draining of blood, morale, and especially money into the continuing Algerian war, which last year's government prophesied would be won by last autumn, and finds, indeed, practically the same government, since the new one set up last week by Radical Socialist Maurice Bourgès-Maunoury is regarded by all as a hand-me-down copy of Socialist Guy Mollet's preceding government, the only difference being that this one is so feeble it cannot possibly live as long and will doubtless die by summer's end, maybe sooner. To make the verisimilitude painfully perfect, the incoming American traveller, in full tourist season, will also find the French Line ships, such as the elegant, elderly Ile-de-France and the giant Liberté, tied up like houseboats at the docks of Le Havre because of a strike, just the way the line's boats were tied up before Christmas in the same kind of strike. Paris newspapers, perhaps through mixed embarrassment and national pride, have relegated this news to their back pages. Yet it is the most important, if unwelcome, news of the week for French shop- and hotelkeepers, and even for the new Minister of Finance. Any interruption in the incoming crop of American tourists obviously means a cut in that richest financial harvest of all just at a moment when the French state, facing an extremely acute monetary crisis, is already borrowing against the Banque de France's gold reserve; has been forced to restore import quotas on goods bought from the member nations of the Organization for European Economic Coöperation, in order to help close the foreign-exchange leak that has been sapping France's economy; possesses the biggest dollar deficit of any member of the European Payments Union; and is literally in need of every red Indian cent she can garner. The Havre strike is of the new type, in that it is not a strike of the common seamen but of the marine gentry—the engineer officers, who are demanding a pay increase of from ten to fifteen per cent, which would make their chief's salary equal to that of the ship's captain himself, in an imbalance of authority, privilege, and reward never before heard of since men first took to the sea in ships. As may be recalled by passengers immobilized aboard the Flandre at Le Havre in last December's strike, the striking engineers then described their demand as "a new philosophy of the modern sea," according to which the highly educated technicians necessary for the operation of the complex machines that nowadays make boats go appear to have taken over the arcane mystique of Aeolus and his winds, as masters of power. They explained that they regard seagoing today as a responsibility divided in two—half to their chief below deck, seeing to the boat's mechanized innards, and the other half to the captain above deck. This would seem to leave a French marine commandant merely the captain of his soul, rather than of his ship.

Though Bourgès-Maunoury's government is called a stand-in for Mollet's government, both could properly be called Robert Lacoste's government, he figuring in each as the all-powerful Resident Minister in Algeria, the dominating chief of the Algerian war policy. Amid France's present high industrial prosperity and unprecedented commercial expansion, the war, which is costing a billion francs a day, has been the plague bleeding the French economy white. To pump up new financial blood, Bourgès-Maunoury will immediately ask Parliament for, and doubtless obtain, the high new emergency taxes that, when Mollet asked for them, led to his downfall. These will feature a ten-per-cent increase in corporation taxes, a jump in postal, telephone, and telegraph rates, an increase to twenty-five per cent in the sales tax on luxury goods, and an alarming boost in the price of gasoline. French gas at the present seventy-six francs a litre is already dearer than the gas of any other European country. The proposed new price of ninety francs would make it equivalent to almost a dollar a gallon. Bourgès-

Maunoury will also free gas from ticket rationing, which France alone has maintained ever since the French and English invasion of Suez. The first French ship since that fiasco and France's subsequent boycott of the Canal is about to pass through the waterway, her toll to be paid in sterling directly to Nasser's Canal administration, which refuses francs—a double, bitter piece of humble pie for France. Ironically enough, Bourgès-Maunoury, as Minister of National Defense in the Mollet Cabinet, was *l'homme de Suez* last autumn, the political leader of that ill-fated adventure. The only novelty in his new Cabinet is a Minister of the Sahara (also a Mollet idea), who is considered necessary because of France's belated recent discovery of rich Saharan oil deposits at Edjelé, Hassi Messaoud, and Ouargla. They are all in Algeria, which only intensifies the dominance of the Algerian question politically, economically, psychologically, ethnically, and morally.

Unfortunately, Lacoste has taken such a harshly repressive Ministerial stand against certain journalists and their writings—among them Jean-Jacques Servan-Schreiber, of *L'Express,* whom he caused to be prosecuted as a demoralizer of the French war aim—that few clear reports, criticisms, or analyses of the realities of the Algerian problem have been printed until this week, when a small brochure called "La Tragédie Algérienne" appeared. It is from the pen of Raymond Aron, the brilliant, nonpolitical staff writer for the conservative *Figaro*—which immediately expressed its disagreement with his booklet. He opens by quoting Montesquieu's austere words: "Every citizen is obliged to die for his country but not to lie for it." He then tells what many Frenchmen either fear or hope has long been the truth—that it is to France's interest to give up Algeria and recognize what he calls "Algeria's vocation for independence," and that it is better to deal now with the rebels of the National Liberation Front than to try to exterminate its men by the hundreds of thousands in a so-called war of pacification and then deal amiably with the remains of the population. Recognizing that France feels she is also fighting for her national honor, he points out that "the maintenance of unconditional French sovereignty is impossible," because it is seeking to break the will of another people. He concedes that the examples given by Tunisia and Morocco since their independence "are not encouraging," but maintains that "to send an army of four hundred thousand into Algeria is senseless" after freedom has been accorded to the two other North African French holdings, and that there should be a logical policy to cover all three. Since Saharan oil offers the first bright economic prospect Algeria has ever had, though it can hardly offset the country's "barren underdevelopment by the French," he says, "the richer the Sahara, the more necessary it is to come to an understanding with the Algerians." He adds, "The best way for France to lose the Saharan oil is to want to keep it for herself." His solemn final lines are "The nationalist demands, with their mélange of religious and racial fanaticism, of Western ideology of self-government, and of the aspiration of humanity toward equality, are a fact that cannot be ignored without a catastrophe. The grandeur of power is something that France no longer possesses, can no longer possess."

July 11

By a generous majority, the French National Assembly, after a debating session that lasted all day and all night, has just recommended the ratification of a plan new in Europe's history—the European common market, aimed at uniting the six democratic nations of France, Italy, West Germany, the Netherlands, Belgium, and Luxembourg in a nondiscriminatory trade community.

What is clear is that three years after the French parliament and people rejected the European Defense Community (France's own invention), which dealt federally with armies levied against possible war—always the first thing European men think of—the common market repeats the federated idea but applies it to business and money-making, civilized mankind's second preoccupation. Whatever idealism there may be in the common market is, fortunately, worldly and hardheaded. In the Assembly debate, the government's State Secretary for European Affairs put his finger like a compass needle on Western Europe's necessary direction for survival by saying, "We are still living on the fiction of the four great powers [of which France was one]. In reality, there are only two—America and Russia. Tomorrow there will be a third—China. It depends upon you," he warned the deputies, "whether there is a fourth—Europe. If you fail to make this choice, you condemn yourselves to walking backward toward the future."

Among the ways the common market will affect the common

man is by letting a citizen from any of the six countries go into any of the five others to hunt work, to shop, to found a business, or to set himself up in his trade—all things that are close to impossible now. This freewheeling of men was the most difficult aspect for the French to swallow, knowing that no Frenchman would set up anything in Germany but that hard-working Germans might successfully set up all kinds of things in France. As one deputy said, "Better a dialogue with Germany than a monologue"—that is, letting Germany listen only to her own voice, as in the fatal past.

In the Assembly vote, the three so-called Old Fathers of the United Europe idea—Schuman, Reynaud, and Pleven—spoke in its favor with passion, hope, and experience. The remnant of the Gaullist party voted against, their diehards and General de Gaulle himself still believing that France should stand on her own, on glory, with no surrender of her nationalism. Mendès-France and his handful of followers (from whose leadership he has just resigned) also voted against, probably for the same reasons that led him to help defeat the E.D.C. when he was Premier—his conviction that England should do her duty and join in to make Little Europe valid and larger. One of France's particular tragedies today is that both these men—the transcendental, Gothic general and the brilliant, Oriental-faced politician, the only new leaders and doers since the Liberation toward whom, for a time, the French people have instinctively turned with belief, aspiration, and a stirring of the sense of security—should now, when they are most needed again, each in his separate way, both be absentees in retirement, though still part of the vital furnishings of France, like the largest, most important portraits hanging on her walls.

The trouble with most new operas today is that it is the melodic old operas that go on being popular. The Paris musical event of this year was the recent first night, at the Opéra, of Francis Poulenc's "Dialogues des Carmélites." It was ordered as an original opus by La Scala, which gave it its world première, in Italian, last January, when the notoriously difficult Milanese critics and gallery gods received it enthusiastically, and when it impressed the major French critics, who all went down, as something long desired and new—a popular success in the field of contemporary opera. Reportedly, the Italians gave it the works—overelaborate scenery for the Carmelite nunnery, and singers with gorgeous voices singing in their great open, emotional, lyric style. Since Poulenc, who was present, felt that the Italian treatment was too worldly and operatic for his pious intentions, the recent Paris presentation became, instead, practically a miracle of impressionistic, controlled musical nuances, echoing in correct gray stone Gothic décors. The French critics who saw and heard both versions found them different, almost opposed, works. Poulenc dedicated his score to Debussy, who "gave me the taste for writing music," and to Monteverdi, Verdi, and Moussorgsky, "who have served me here as models"—which is occasionally true, especially of two or three touches of "Boris." His opera score is marked by his characteristic ecstasy of expression and subtlety of harmony; by lofty reaches of mounting melodic grace; by the rich polyphony of his chorals, such as the "Salve Regina" and "Ave Maria"; and by an all too brief last-act overture of really passionate loveliness. The opera is too long. This lets Poulenc's talent seem to *manquer de souffle*—to run out of breath. Nor were the Paris Opéra voices all gorgeous, by a long shot.

Most operas feature trouble between the sexes. This opera, about chaste nuns and with almost no male characters, is taken textually from the play "Dialogues des Carmélites," presented here five years ago—a tremendous dramatic success written by the late, highly intelligent Georges Bernanos, who, in turn, took it from a German novel, "Last on the Scaffold," which is indeed its *Hauptsache*. Its heroine, Blanche de la Force, is a purely fictional aristocrat who has sought the cloister because she is morbidly afraid of the violence of life, yet, in imitation of Christ's agony, finally mounts the French Revolutionary scaffold as a volunteer, the last to die among her condemned Carmelite sisters (real historical characters these, beatified by the Church in 1906 and now also part of the opera). Its final scene shows the sanguinary mob before the offstage scaffold, as a bloodcurdling offstage mechanical noise imitates the repeated falling sound of the guillotine's knife. Poulenc's opera has already been given in Germany, where it opened in Cologne last week, and is listed for production in San Francisco.

The ten-day European heat wave, with the Paris thermometers standing officially at ninety-seven degrees in the afternoon and actually at a hundred and twenty on midtown sunny balconies, seems now broken by storms. Exceptionally, the Paris police were allowed to shed their woollen dolmans, or jackets, and direct traffic in their pale-blue shirts—very smart with their white revolver holsters and white clubs. Wedding photos of the Comte de Clermont, dauphin pretender to the Bourbon throne of France, showed him mopping the sweat from his royal brow during the church service at Dreux. More than a fourth of the tough professional bicycle riders in the annual international Tour de France simply gave up by the third day. Those who stuck it out wore fresh cabbage leaves under their caps for insulation. When they pedalled through the big cities, the local firemen doused them with fire hoses. All last week, the Paris sidewalk cafés were running out of ice, beer, vanilla ice cream, and unsqueezed lemons by evening. The apéritif favored by the French was Pernod, which, because it is based on aniseed, they benevolently regard as a fortifier of the intestines during heat, and thus a classic thirst quencher. On July 4th, the sun set some tar paper on fire on the roof of the unfinished new UNESCO building, near the Eiffel Tower. Parisians, who by no means stand heat well and who incline to see international politics in everything, called the hot spell "atomic weather," forgetting, until their own weather bureau reminded them of the fact, that it was even hotter here, and for a longer stretch, just ten summers ago, when those terrifying weapons were infants.

August 17

American tourists are giving Paris the go-by because of high prices, and two million Parisian workers and employees are off on paid holiday, so the city is delightfully empty—ideal for a resident on vacation. Owing to the crisis in the falling franc that preceded its devaluation and the fact that paper pulp for newspapers is imported, the Secrétariat d'Etat à l'Information limited the number of Paris newspaper pages to eight during the rest of August, so we are also isolated from all except the most signal news. One such item that just received full space in all the dailies is that five neighborhoods of the city are suffering an invasion of termites—in Passy, around the Gare d'Austerlitz, around the Sorbonne, in Ternes, and in St.-Germain-des-Prés. Apparently, this is the worst invasion since the time of François I (1494–1547)—a pretty clean record.

September 10

M. Georges Salles is leaving the Louvre, where for thirteen years he has been the Directeur des Musées de France and a distinguished, energetic, creative-minded figure in modern European museology. He came to the Louvre in October, 1944, when it was still emptied of most of its treasures—an unheated, desolate, non-functioning edifice—and his gigantic task over the first year was to bring back its great belongings. This writer recalls standing below the Escalier Daru on the day when that enormous antique fragment, the Winged Victory of Samothrace, was being restored, inch by inch, to her former position at the top of the steps—was being slowly slid upstairs on primitive greased wooden runners, the power being supplied by a series of ropes, pulled with appropriate strength, and yet with the right watchful delicacy, by the Louvre workmen's arms—an impressive, old-fashioned demonstration of pulleys, hemp, and expert judgment. Under Salles' aegis, some significant experiments and changes were made in hanging the Louvre's special wealth of European paintings, the aim in each case being aesthetic elegance but also the educational and psychological effect on the public, for whose edification, after all, the art is on view. In 1948, for instance, the school system of picture hanging was reorganized, the Italian school being hung in the Grande Galerie, the Spanish in the Salon Carré, and so on. Then, in 1953, the orientation was completely altered to the satisfying presentation still used today, which is according to the relationship of styles, so that you can see mingled, rather than separate, the chefs-d'œuvre of all the European national schools during the great stylistic periods, and also their regional differences. M. Salles also influenced the Musée d'Art Moderne here by encouraging the postwar Conseil des Musées to buy for it modern masterpieces of the kind that the French state, in its shortsightedness, lacked—seven Matisses, four Braques, three Bonnards, and a Laurens sculpture being purchased in 1945 alone. In 1955, the first exhibition of pre-1850 French art ever seen in Japan was shown in Tokyo. It was organized by Salles and proved to be the greatest stroke for French prestige achieved in the East in recent times—an epoch-making aesthetic event for the Japanese, more than a million of whom attended the show. Modern French art they knew; it was what came before that so excited them, such as medieval stained glass and the pictures of Chardin and the growth of French art in the seventeenth and eighteenth centuries.

September 25

Wednesday morning, the Communist paper *Humanité* front-paged a drawing called "School Opens in Little Rock." It showed Father, Mother, and little son at home enjoying their televised news, which was showing them a lynched Negro boy hanging from a tree. The conservative *Figaro's* special French correspondent in the United States referred in his long Wednesday dispatch to "the ignobilities" of the previous day that had "provoked President Eisenhower to a spectacular dramatic action"—duly announced in a mid-page *Figaro* headline as "DES TROUPES AEROPORTEES ENVOYEES A LITTLE ROCK"—to protect nine young Negroes who were merely trying to go to school. The liberal *Combat* ran a Wednesday editorial called "Eisenhower with His Back to the Wall." It said, in part, that "in an epoch when people of color everywhere are trying to emerge with their own entities, the United States is fortunate," since its Negroes "aspire to nothing more than to have the standing, rights, and dignity of the average American." It added that not only was President Eisenhower's prestige being diminished "by the whites' brutality" but there was risk of injury to America's prestige. "In Washington's bitter struggle with Moscow for the friendship of the Afro-Asian bloc," it went on, clever use of the Arkansas events could "show America to be a land of race hatreds and hypocrisy. The rioters in Little Rock have just rendered a major service to Soviet propaganda."

President Eisenhower's proclamation this week, ordering the Little Rock fomenters of trouble and illegality to "cease and desist," plus his radio speech and his decision to employ the Army to restore calm and enforce the law, has naturally intensified French interest in the Little Rock rebellion. Already its daily developments had been passionately followed here. The photos of the white high-school boys' faces stretched in a rictus of hate, the bowed head of Elizabeth Eckford, the strange, démodé visage of Governor Orval Faubus had all become perfectly familiar to the French by name, identity, and meaning, our bad news having become an international affair to which everybody has paid close attention. Besides, the French have an exceptional knowledge of our Southern scene—in most cases without ever having visited it in the flesh. French children are taught a great deal more about our Civil War than American children are taught about the French Revolution. French intellectuals have a larger knowledge of Southern character, through reading Faulkner, than our American intellectuals have of any French regional class except the *faubourg* group of ducal Guermantes and friends in Marcel Proust. In a character sketch of Faubus, the independent *France-Soir* disparagingly described him as *"un hillbilly, qui sont les plus pauvres et les plus arriérés des Américains de race blanche."*

October 9

The first reactions here to the Soviet Union's launching of its artificial satellite were almost poetic in their felicitously phrased astonishment. Parisian papers called it "a myth become reality" and "the first proved possibility of an effective evasion from our planet," said of it that "the road to the moon has been opened" and "a giant step in the sky has now been taken," and wrote that "for the first time, an object fashioned by the hand of man is promoted to the condition of a celestial body, endowed with an independent existence in cosmic space." Phrasing it more simply for history, one Parisian paper commented, "Friday, October 4, 1957, was the first day of the Year 1 in the interplanetary era." Yet it was not until Saturday morning that the Communist Party paper *Humanité,* like every capitalist journal in Paris, came out with the world-shaking news. You might have thought that for propaganda or prestige value the faithful Comrade paper here would be given a scoop on this astounding, victorious Soviet event of the century. But, as usual, the French Communist Party was treated by Moscow like a poor country cousin. The Tass Agency must have given out its satellite news so late that *Huma's* front page could not be changed beyond squeezing in a last-minute headline above its title, saying "PREMIER SATELLITE ARTIFICIEL LANCÉ HIER PAR L'UNION SOVIETIQUE." On page 2, usually the theatre, movie, and book page, it ran the Tass Radio Moscow story, plus a triumphant little paragraph about the *"vive sensation"* in Washington.

There are no French Sunday papers, so it was not until Monday that *Humanité* swelled into its limited propaganda stride. The Tory *Figaro* of Monday morning carried four pages on the Soviet satellite, plus a huge front-page announcement: "SENSATION THROUGHOUT THE WORLD." But if you think that *Huma* made the Russian savants' achievement its main front-page news, you are dead wrong. Its major headline was about Mollet and Pleven and the French government crisis; football and sports got a big front-page play, and so did the steelworkers' strike. Then came a satellite feature story for the lower-brow echelons, which have already nicknamed it *"le bébé-lune"*; this story was called, believe it or not, "A Baby That Jumps from Bordeaux to Barcelona in One Minute." And, finally, tucked away among this front-page rubbish was a short, official, pure-propaganda editorial, called *"Le Spoutnik"*—a word that it never once translated for its French readers. Its theme, of course, was that this new mastery of celestial nature is owing to Soviet science, and that Communism alone created the social conditions in which Russian science brought to flower the world's first satellite, as well as the world's first intercontinental ballistic missile, whereas "in lands still subjected to capitalism," like the United States, "scientists complain of the insufficient means put at their disposal for research." Then came the sombre warning: "The launching of *le spoutnik* should riddle those Western propaganda lies that seek to make people believe that when the Soviet Union talked about its intercontinental missiles it was only bluffing." This was followed by the carefully baited illogical question "Does not the launching of this satellite prove that science today has reached a stage where a happy life for all mankind can be assured only if it is relieved of the burden of armaments?" And "Mankind is now prepared to conquer the heavens" was the editorial's conclusive, high answering thought, which is probably the truth, too.

There has naturally been some shrill French laughter at the poor astronautic position that the United States finds itself in—especially in this International Geophysical Year—with the Russian-manufactured baby moon beep-beeping over Paris, Washington, and New York as it hurries around the world fifteen times a day, just after our Atlas rockets, supposed to hold up the defense framework of the Western world, fizzled out and fell flat. It is constantly being remarked by Parisians that America talked about—indeed, is still talking about—its satellite, but that the Russians built and launched theirs, which is the one being talked about right now. There is a folkloric phrase being much cited here—"selling the bear's skin"—which the French find especially suited to the American satellite situation. The phrase comes from a fable about a braggart who sold a bear's skin before he killed the bear.

France still has no government. This makes everything seem nice and quiet. M. Guy Mollet, who was Premier last year, was asked last week if he wouldn't be Premier again, and he said yes, but the deputies said no, and refused to have him. Then M. Pleven was asked, and he said yes, but the deputies said no once more. Now M. Mollet has been asked a second time, and has said no. Pretty soon, somebody will say yes, and the deputies will say yes, too. Then France will have another government and the dissensions and rigmaroles can start all over again. It is something to look forward to.

October 24

This has been a worrying, busy week—the week of France's most overdue engagements with her immediate 1957 history; the week of unavoidable dates with her people, her politicians, and her bank account; the week of finally facing up to recent reality, on which her Fourth Republic must try to survive. On Monday, the twenty-first day that France was without a government, the week's vital opening date was with the experienced seventy-one-year-old M. Robert Schuman, of the Catholic Mouvement Républicain Populaire, who, it was hoped, might be allowed to become the new Premier, three previous candidates having failed, but before he considered trying to head a new regime, he called on M. Wilfrid Baumgartner, governor of the Banque de France, from which the state has already borrowed billions. A former Ministre des Finances himself, Schuman reported succinctly to President Coty that "France's financial situation is tragic." His second information for Coty was that until the left of center, which is Socialist, and the right of center, which is moderate, pulled closer together in their notion of what might save France, nobody could found a government at all, though clearly one of some sort is initially necessary for her salvation. Even possible national bankruptcy is apparently a pleasanter prospect for most deputies than the one vainly proposed a few days ago by the conservative Premier-candidate Pinay, who dared demand, as one basis of his program, that for an entire year he be allowed to govern without their interference.

This present difficult week, openly called *une semaine d'agitation sociale,* began with what was also called extreme tension. After the general public's disgust on Monday that there was still no government (which, once in, it always regards with aversion), labor troubles became intensified. The unexpected amplitude and success of the previous week's strike of the men in the nationalized gas and electrical companies, which practically paralyzed Paris and France for twelve hours, were naturally heartening, surprising, and inspiring to French workers and employees, cruelly squeezed of late between constantly spiralling costs of living and, in most cases, the same old immutable pay, and further harassed by the deputies' inability to get their own work done, which was to set up the next regime. On Monday, plans were announced to make Friday of this week "a great day of action, pay demands, and protests," most notably by a major, crippling twenty-four-hour strike of the French railways, beginning at 4 A.M., accompanied by a strike of the same duration by the postal and telegraph services. Other strikes scheduled for Friday are of the building trades, the metalworkers and the synthetic-textile workers, the Métro and Paris bus workers, and, above all, the vast personnel of the state civil service and the governmental offices, including that legion of scribblers in the tax bureaus—plus a partial strike among schoolteachers, who are to refuse to teach classes of more than forty children in their overcrowded schools. Latching on to the band wagon, even the policemen's union urged its members to avail themselves of the big day "with the only legal means at their disposal"; i.e., a mere request that their monthly salary be raised by seven thousand francs, it being legal for them to ask for it but not to strike for it. During the week, the unrest has been manifested in big, practiced proletarian demonstrations, such as walkouts by the steelworkers in the Loire Valley, and in Nantes and St.-Nazaire on the Atlantic coast, and in smaller gestures by the most refined artisans, such as the tailors at Lanvin and Creed, reported to be laying down their needles for a fifteen-minute strike every hour.

Though the Communist-dominated Confédération Générale du Travail, the biggest and toughest union, has naturally been bossing the Friday plans (which the Socialist Force Ouvrière has rather snobbishly declined to participate in on equal terms), it is the reaction of usually the mildest of the three unions—the Catholic Confédération Française des Travailleurs Chrétiens—that has furnished the most astonishing new labor element. Not only has it backed the C.G.T.'s Friday plans but it has boldly invited its followers "to give the day of October 25th the general character of a warning to the state and to the employer class," and, further, "to give an expression of the determination of the working class to say no to the rising prices, to the lowering of living standards, and to social regression."

The financial experts have been talking gravely during the week of the effect that the important increase in wages being demanded by the strikers will have on France's perilous inflation. Yet none of the major labor groups or unions so far is asking for wages that match the increase in the price of turnips, which cost forty per cent more than they did last year at this time, or string beans, which are sixty per cent higher, or tomatoes, which have gone up sixty-six per cent. Now Socialist Guy Mollet is again going to try to be Premier of a government—of limited life, he says. Since the war, France has only too frequently lived in the limbo of a protracted political crisis without a government. But this is an extra-grave political crisis, currently in its fourth week, with barely a mirage of a government in sight, with victuals dangerously expensive, with massive labor troubles about pay accumulating amid an alarming financial situation and amid the persistence of an insoluble, continuous, and costly colonial war.

There has been general, but nothing like complete, satisfaction here over the fact that a French writer was awarded the Prix Nobel de Littérature and that this Frenchman was Albert Camus. For there are a great many readers and writers who agreed with Camus himself when he said, with his insistent truthfulness, "Had I been on the Swedish jury, I would have voted for Malraux." In making the award in the office of Camus' publisher, Gallimard, the Swedish Ambassador compared him to the resistant seventeenth-century heroes of Corneille. In thanking him, Camus significantly mentioned his appreciation that such a prize had come to "a Frenchman of Algeria"—a mere colonial. Young for the honor, though twice married and with adolescent boy and girl twins, he was born in November, 1913, near Bône, of humble farm laborers, his father being killed the next year in the Battle of the Marne. Poverty and the maladive beginnings of what finally became tuberculosis interrupted his studies in philosophy (partly on St. Augustine), and he had to earn his living at odd jobs—in a garage, as a shipping clerk, as a schoolteacher, and, finally, as a journalist, first in Algiers and then in Paris. During the war, he joined the Resistance group called Combat, whose little clandestine newspaper became the postwar intelligentsia Paris daily *Combat,* and after the Liberation he worked on it as an editor, along with Sartre and Simone de Beauvoir. His literary career had started during the Occupation with the publication, in 1942, of "L'Etranger," the first of his novels, which, in retrospect, the French oddly see as American in their sparse, unheroic style but utterly Mediterranean in thought, with the antique, abstract qualities indigenous to that early, educated, civilizing region. In "L'Etranger," the man Meursault becomes the figure of truth, as a form of nihilism. In his philosophical essay "Le Mythe de Sisyphe," man's vain, constant push toward the summits becomes not punishment but stoicism, a repeated form of moral courage. "La Peste," published in 1947, is his most famous, influential novel on the theme of good and evil, which nobly obsesses him, and to the French its two most illuminating statements are still "What is neutral is the microbe" (of the mysterious plague), and the lofty, agnostic query of one of the men fighting it—"Can one be a saint without God? This is the only concrete problem that I know of today." Camus became and remains for the French what Mauriac has called "the conscience of his generation"—the spokesman for the outsider in today's moral crisis of alarm, anger, and despair, and for that rather young war-raddled European generation, freed from religion but still the captive of its ceaseless, puzzling questions about man on earth, to which Christian faith formerly supplied all the answers. In 1951, his essay "L'Homme Révolté," the most controversial of his writings and the hardest to read, analyzed man's instinct for revolution, a subject dear to French hearts, and concluded that revolutions always create their own tyranny. This and his ideas on Soviet labor camps led the following year to a break between Camus, on the one hand, and Sartre and pro-Communism and Existentialism, on the other—a public quarrel much enjoyed in Paris because the men fought it out in Sartre's magazine, *Les Temps Modernes.* "La Chute," or "The Fall," his latest novelette, which is a fascinating study of justice gone mad, will be followed by one called "Le Premier Homme," which he is now writing. Since youth, Camus has worked around the theatre—has run his own troupe, has written and adapted plays, and has just finished an adaptation of Dostoevski's "The Possessed," which takes four hours to perform and will unquestionably be played here even so. Camus is certainly the only obsessed moralist the Swedes could have found among the French today to receive a literary prize for "earnest study of the problems of the human conscience in our times."

November 5

This crisis of thirty-seven days without a government has left its mark. Not since the wave of anti-parliamentarianism that culminated in the bloody riot of 1934 on the Place de la Concorde has there been such openly expressed disgust among the French for their National Assembly as a weak, selfish, do-nothing institution. During the crisis, the French did not even want to talk politics, as if politics were a sort of incurable malady raising a stench over the land. At the movies, whenever deputies were shown in the newsreels the audiences, including many of the women, booed and jeered, and the younger men usually shouted a few insults. But that was all. After the four-year occupation by the Nazis and the French people's bitter disillusion with their Fourth Republic since then, there is no climate for rioting. Even the faint rumor of a Fifth Republic, under the magic banner of de Gaulle, has aroused nothing but derision and disbelief. This last week, the French deputies refused (for the moment, anyhow) to accept an increase in salary, now being offered to senior civil servants generally, which would have given them an additional seventeen hundred dollars annually. Prudently enough, they thought parliamentary prestige was so low that their higher pay would only disgust their electors further.

November 20

The Allies' intergovernmental relations are usually on such a talkative high level that the ordinary citizen here makes little effort to follow their ins and outs, but this certainly was not the case with the British and American delivery of arms last week to President Habib Bourguiba, of Tunisia. Since, for once, this was a matter of acts, not words (until France's new Premier and her ambassadors, politicians, radio commentators, and newspaper editorial writers let loose a thesaurus of indignant phrases), the French people felt that they understood it perfectly—that it was a treacherous, inimical gift to France's ex-rebels of guns that would be easy to pass over the border to her present rebels in Algeria. Parisians, in their talk on the subject, are roused and galled, and forgetful of Premier Gaillard's levelheaded parliamentary statement on Friday that London and Washington had warned Paris they would have to send the requested arms, since France had refused to and Moscow and Cairo would be delighted to oblige. Parisians, however, remembered his other statement: "If the Atlantic Pact should fall to dust one day, we will know the artisans of its failure"—John Bull and Uncle Sam, not La Belle Insouciante Marianne. So acute was the tension that on Friday and Saturday nights, five hundred police from the riot squad, wearing steel helmets, lined up before the American Embassy, just off the Place de la Concorde—unnecessarily, it turned out—and similar unneeded protection was given to Her Britannic Majesty's Embassy, and the Embassies of Tunisia and Morocco. The American Embassy says that by Monday morning it had received an "avalanche" of letters from parents of French soldiers in Algeria asking that they no longer be sent *Information et Documents,* a free and till now popular biweekly magazine about America published in French by the United States Information Service. The parents also expressed their anger, shock, and horror at the thought that the United States should have sent arms to North Africa, where they could be handed over to Algerian rebels to kill their boys.

America's already well-established unpopularity here progressed over the weekend by leaps and bounds. The arms-for-Tunisia incident, coming, as it did, after the recent discovery of rich oil fields in the Sahara, led to a belief that this time the Americans had become disloyal to their ally in order to satisfy their cupidity. The always excitedly nationalist paper *L'Aurore,* strongly Rightist in its views, said editorially, "The oily explanation [*l'explication pétrolière*] of the Tunis maneuver is only too obvious." The conservative *Figaro* called the arms shipment "an odious blow by our allies." The august *Monde* referred to "the entente without cordiality." Ex-Premier Georges Bidault protested, "We cannot be the ally in Europe and the scapegoat in Africa." Raymond Aron, a noted non-Leftist political writer (who is, notwithstanding, anathema to the Right Wing, because of his belief that Algerian independence is practically inevitable), voiced his distress that London and Washington should have "chosen such a deplorably clumsy method of expressing no confidence in France's Algerian policy." Ex-Ambassador André François-Poncet, who until recently was stationed in Bonn, was unique in publicly expressing a mixture of Gallic indignation and diplomatic wisdom. He said that the arms shipment was "not only inimical but also shocking and humiliating, and marked by the brutality that, it seems, M. Foster Dulles has an increasing tendency to utilize." The Tunisian incident, he continued, "without doubt expressed our partners' impatience with and disapprobation of our country's politics, our quarrels, our interminable Ministerial crises. The lesson they intended to give us is not absolutely unmerited; it would be unmanly not to admit this. But the Anglo-Saxons made a bad calculation. They dealt the Atlantic Alliance a disastrous blow."

November has been a month of anniversaries for the French, nearly all of them disagreeable. On November 1st, three years ago, the Algerian war began, as a supposedly unimportant native uprising. On November 5th of last year, the French and British vainly sent off paratroopers to blast Colonel Nasser out of the Suez Canal. On November 7th of this year, Moscow celebrated the Bolshevik Revolution of forty years ago, which began the era of Soviet Russia, whose theories a fourth of the French Republic's voters supported in the last election. And this November 5th was the thirty-eighth birthday of Félix Gaillard, and the day on which he took office as the seventeenth and youngest Premier of the Fourth Republic. He became Premier not because, being young and brilliant, which he undoubtedly is, he stood for a rising, intelligent, new generation of leaders in stale French affairs but because, as the French said only too clearly, the old parliamentary hands' selfish disunion was more precious to them, as a major part of politics, than getting together and governing France. It is thought here that possibly the one ameliorating result of the Tunisia-arms incident is that Premier Gaillard will not be allowed to fall for a while. There will, of course, be plenty of time for that later.

As a natural result of social unrest, government financial troubles, inflationary prices of consumer goods, food, and *vin ordin-*

aire (they have just gone up for the third time since early autumn) and the greatest industrial prosperity and bourgeois spending boom that modern France has known, there has been a rain of strikes by undersalaried employees, especially those who have the honor of working for the City of Paris and the French state. The latest strike of *petits fonctionnaires* took place on Tuesday. The strikers' demand was for an increase in the basic wage on which all salaries in the lower level of government services are calculated. The clinics for diagnosis and treatment in the city hospitals were closed, garbage was uncollected, gravedigging was suspended, and penitentiary guards quit their jobs. During the one-thirty radio news broadcast, the speaker announced that the station would be off the air until eight that evening. The zoo and the public libraries, including the Bibliothèque Nationale, were closed; water and gas pressure was low. The meteorological services struck, and so did the customs officers; the Paris airfields were shut down; Air France cancelled all its European flights; long-distance phone calls to points outside France were nonexistent. However, letters were mostly delivered, and in the general confusion telegrams were usually distributed twice.

This has been a great month for one pleasure, at least—the splendid art shows customary at this time of year. Finest and rarest in content of them all is the exhibition called "L'Atelier de Juan Gris," in the elegant new Rue de Monceau gallery of that early Cubist-art connoisseur Daniel-Henry Kahnweiler and his sister-in-law, Mme. Louise Leiris. Gris died in 1927, only forty years old. The twenty-two canvases now displayed were painted in the last year, and even the last few months, of his life, and were found after his death in his studio in Boulogne-sur-Seine, a suburb of Paris. They had never been seen by the public.

When Gris died, his small pictures were selling, if he was lucky enough to find a buyer, for a few hundred francs apiece—perhaps no more than twenty dollars in those days. Their sale price in the present show is six million francs, or around fourteen thousand dollars, and the big ones get more than twice that. During his brief lifetime, Gris was appreciated only by a very few, and their number did not include his compatriot Pablo Picasso, whom Miss Gertrude Stein bluntly reproached for his phrases of what she said was false grief after Gris's funeral. These twenty-two canvases, the product of his last burst of genius, have the pristine freshness, in color, of paintings that look new because they have been hidden away, have never been handled or exposed to light and public opinion. There is in their limpid yet austere composition a reflection of the certitude of Gris, which here has a meditative quality that came perhaps from his sense of approaching death and his final sureness in his art. The largest still-life, "Guitare et Papier et Musique," offers as a startling surprise, considering how low his palette usually was, a patterned blaze of vermilion, as red as blood. Two paintings, each showing a large seated female figure of neoclassic dignity, are also rarities. One of the figures holds a harvest basket of fruit, the other (one of three pictures in the exhibit that are signed) has empty, folded arms and a strange, graceful, gray lunar face. Of his small final *natures mortes,* one depicts a carafe and lemon, another a wineglass and egg, and there is one with bananas. The small, perfectly produced catalogue that M. Kahnweiler had printed, surely at considerable expense, is a collector's item. Its cover, which bears a stunning colored lithograph of the canvas called "Fruits et Bol," was prepared by Mourlot Frères, the great lithographers of Paris.

The *clou* of the Salon d'Automne—in the gloomy, clammy Grand Palais, which is probably the worst place on earth to show art, but which in its time has shown the beginnings of the great Ecole de Paris, and the last, in posthumous retrospectives, of Cézanne and Gauguin—is a retrospective show of about fifty canvases, dated from 1900 to today, by that belatedly appreciated master Jacques Villon, now aged eighty-two. He is still painting with the purity of research that he pursued during those long, too quiet years when he was unjustly ignored. The exhibit traces his aesthetic history from his early, almost banal period of youthful realism, through his Cubist stylizations, to the refined geometric patterns of color, like shards of rainbows, that have become the style of his present apotheosis.

The chill of November has finally arrived, after half a month of Indian-summer sunshine and a full complement of yellow leaves—comforts left over from October. Not for years has the autumn season been so extended, mild, and handsome in the city gardens and in the forests around Paris. The wild-mushroom harvest in the woods was particularly plentiful. Morning mists, so heavy that they dripped moisture onto the leaves and from there to the ground, and then warm afternoons, as the sun burned the mists away, gave all the fungi their ideal climate. In certain woods in the Ile-de-France, the crop of *les trompettes de la mort,* or *Craterellus cornucopioides*—the so-called trumpets of death, though no mushroom hereabouts is safer and tastier—was especially heavy and succulent. *Les trompettes* usually grow beneath young oak trees, in clean glades devoid of underbrush, rising from cushions of green moss, where they congregate in clusters, looking like purple, black-shaded crocuses. Floral-shaped, tubular, decorative, mysterious, they are difficult to find, half hidden by last year's fallen leaves. In taste similar to the dark morels of spring, they are autumn's richest gift of all such spored growths here. If they are to be tender, they should be cooked until they turn as black as dead flowers. They are then served with rice *al dente* and a little olive oil, and consumed with a gourmet's appreciation, and no fear, since for once nature has not created any deadly duplicate. Ten days ago, there were still a few handfuls of red raspberries in the gardens, and the small wild strawberries planted by the garden paths were again in bloom, uselessly. Now winter has started to settle down, and with a rather special melancholy, owing to turbulent human events.

December 4

The present overinflated art prices, demonstrating the cheapness of today's money and the exaggerated social or investment value placed on modern canvases, have naturally resulted in a lively wave of counterfeits of contemporary French masters. The French police recently nipped in the bud a plan to sell a bundle of fake Picassos, Braques, and Utrillos, among other counterfeits, in Texas. The situation has led to an investigation here of the experiences that have befallen the descendants of some of the dead great French painters, which are both interesting and alarming. Mme. Cachin-Signac, daughter of the Pointillist, says that anything styled in little dots tends to become a Seurat or a Signac, but that occasionally, with permission from the owner, and in the presence of a bailiff, she has been permitted her right—that of scratching her father's signature from a false canvas. Isabelle Rouault, the painter's daughter, says that with the permission of the owner she has sometimes been able to deposit a fake Rouault with the Syndicat de la Propriété Artistique, which prevents its being circulated, and has even succeeded in destroying some bogus canvases. Pierre Cézanne, himself a painter in Montparnasse, says that he has seen hundreds of false pictures attributed to his grandfather; that the owner of an exposed fake is always furious with him; that lately he twice saw the same false Cézanne, which the second time contained a couple of little houses newly painted into one corner of the canvas; and that most of the fake Cézannes are copies of works done between 1870 and 1880, and are invariably signed—often the real evidence of their falsity, "since the signature of my grandfather during those years was rare."

1958

January 8

The publication shortly before the year's end of André Malraux's latest illuminating illustrated book on art, "La Métamorphose des Dieux," put it in a correct position to be the most notable art volume both of 1957, when it came out, and of 1958, when people can take the necessary time to read and ingest its four hundred pages. They are less difficult to read than his six hundred and fifty pages of "Les Voix du Silence," of 1951, provided the reader has actually read that preceding crucible work and is therefore familiar with Malraux's molten ideas, with his glowing memory for the light of art all over the world, and with his fiery literary style. The compliment to him as the accepted remaining genius of French letters since the death of Valéry and Gide was manifested by the really majestic treatment that Paris gave "The Metamorphosis of the Gods"—a treatment of such grandeur and amplitude as no one recalls ever seeing accorded an author before, including Malraux for his earlier books. The first edition of "La Métamorphose," which consisted of ten thousand numbered copies, priced at six thousand francs a copy (more than fourteen dollars, and considered costly here), was immediately sold out—a rare occurrence for so de-luxe and hermetic a work.

As the title of the new book indicates (a second volume will follow), it is a special amplification of the aesthetic theory that Malraux has already stated and has held since youth, when he began his intellectual concentration on art as the earthly immortality of man. This is that "metamorphosis . . . is a law governing the life of every work of art," meaning simply that art takes successive forms in remaining itself over the passage of time—a rationalistic modern notion of universality, which was not shared, Malraux points out, by the nineteenth-century artist Delacroix, to whom Egyptian tombs were not art "but high-class curiosities," and whose ignorance of Romanesque art (regarded for five centuries by the French as barbaric) was "equal to that of Baudelaire, and even Cézanne." Today, however, Malraux declares in a clarion phrase, "From the history of dead civilizations, as from the ethnography of peoples who are dying, we expect to be informed of what man was when he did not look like us."

The body of the book is divided into two main considerations of art—that of the divine and that of faith, or, in other words, the pagan and the Christian. Fortunately for the reviewer, Malraux has provided, in trenchant phrases at the book's end, an analytical table of his ideas, which, since one travels at vertiginous speed in company with his mind, serves in the manner of a railway timetable (as one French critic has gratefully pointed out), so the reader can know which way on earth Malraux is headed. He starts by declaring that the Western world might have known more about the profound questions posed by art's existence if it had not erred for centuries in thinking that the origin of all art was the Greek conquest of the human likeness, and, moreover, in thinking that "Greek sculptors wished the gods to look like men," when their aim was the other way around. Beauty was a matter of divinity, "which set the goddesses apart from mortal women." The statues of the prize athletes—for, as Malraux points out, Greece also invented glory—were "not portraits but ex-votos in the grand style," given by the crowned athletes to the temples, as, later, the Christian princes gave statues of saints to the churches. And when the Greek heritage passed to Rome, in that city, "for the first time, a major art recognized the system of resemblances as the system of the world; for the first time, resemblance became *the real.*"

Malraux's chapters on the art of faith, which was Christian, open with his statement that to replace the former pagan art peopled by statues, Byzantium created a population of immobile apparitions—the holy figures of the mosaics. Then follow chapters illustrated with reproductions of extreme interest to any travellers familiar with the carved Romanesque and Gothic figures of the great churches. As the Gothic style and the veneration of the Virgin flowered, and love, pity, and charity became the new ideal of human faith, the carved life of the Virgin came to settle at home in the cathedrals—Queen of the feudal hierarchy of peasants, seigneurs, emperors, and popes who made up the medieval civilization of faith. In its wake, as an extension, came the art of painting—first in two dimensions, flatly stating the world of God, then (with Giotto) in three dimensions, in loving imitation of man and nature. And far to the north, in Flanders, where painting was also strong, came portraiture and the world of Flemish art. There this volume ends, almost abruptly, with the art of van Eyck, who painted Eve, the saints, and the Virgin as people in a living world, enriched by the reality before the painter's eyes, the reality that he saw existed, whereas in the south Italy was getting ready "to paint Venus because she did not exist."

February 5

President Eisenhower's post-midnight announcement of Friday's successful launching—at last—of our first heavenly satellite, Explorer, came too late for the Saturday-morning newspapers in France, so the Paris *Herald Tribune,* in order to spread the good news through the American colony over the weekend, brought out its Monday paper on Sunday morning (a day when no papers are normally published here by anybody), calling it, rightly, a special edition and dating it Sunday-Monday, February 2–3. It carried the New York Sunday edition's eagle-cry editorial about how quickly America can overtake a lead "once the nation rolls up its sleeves and tackles the job," and it also gave the complete, astonishing details (many already published in the Saturday-afternoon French papers) concerning the former German, Dr. Wernher von Braun, and his ex-German Army group of rocket specialists from Peenemünde, and how it was these captured brains with the rolled-up sleeves who had sent Explorer into its orbit on advanced versions of the V-2 rockets that they had earlier devised and launched for Hitler. These facts, with their full significance of German, rather than American, know-how, cut the enthusiastic edge off the French people's suddenly revived belief in Yankee technical leadership—at least, if one can judge from the disappointed way many Parisians talked, as if they had been sold a false bill of goods. The fact that the Russians also had their parcel of captured Nazi technicians was sufficiently appreciated here a full month ago to make popular a joke about the Russian sputnik and the American satellite talking German to each other in the sky—provided the American satellite ever got up.

March 4

For the first time in history, the little-known opera house in the Palais de Versailles is today open for ordinary people to go in and look at, though few seem to know this, since there are neither descriptive catalogues nor even postcards of it at the Palace stationer's counter to publicize the new privilege, and since only a scant announcement has ever been made in the Paris press about the reopening of this hitherto dilapidated architectural treasure. In the time of the aging Louis XV, when it was finally finished, it was loyally and royally called the most beautiful theatre in Europe. It is still unsurpassed, no other old theatre being comparable in allure, extravagance, and elegance, except perhaps those two classics—the Fenice, in Venice, and the Margraves' private rococo theatre in Bayreuth, both of which benefit in delicacy of effect from the refined fading of their original décors. The Versailles Opéra, which has just cost the Fourth Republic nine hundred million francs and five years to restore, is now as brilliant and as magnificent as it was in the King's day, so that we see it fresh, as if with his protuberant, carnal royal eyes. Its opening performance, in 1770, of Lully's opera "Persée," was in celebration of the marriage of his fatally unfortunate grandson, who was to become Louis XVI, to the unlucky, unpopular Austrian bride, Marie Antoinette.

Apparently, the court architect Jacques Ange Gabriel (he who designed those great pillared Place de la Concorde façades) had a hard time of it after his first plans for the theatre were submitted in 1748, with work and funds typically interrupted by the Seven Years' War and the King's continuing shortage of money for the next twenty years. However, from Gabriel's long, often discouraged concentration came a novelty of technique, and even of orderly ornate beauty. Theretofore, theatres had customarily been rectangular. He conceived and built the Versailles Opéra as an oval of fine wood, with the accurate acoustics of a violin for the ears and, for the eyes, an opulent mélange of green-painted false-marble panels, white fluted pillars, intervening golden statues of goddesses, small golden zodiac signs, and golden balustrades around the lower balconies, topped with bull's-eye windows covered with golden grilles, like elegant prison windows. Rows of benches (now in place again with their blue cut-velvet cushions) served as orchestra seats. The Royal Family sat in state on front-row chairs in the raised dress circle to the theatre's rear, and over their heads was a series of loges, with square golden grilles, behind which they could retire to look at the performance without being stared at themselves. In the King's time, the floor was covered with Canadian bearskins against the cold—replaced today by a modest animal-brown wool carpet. It seems that during the wedding celebration to honor Marie Antoinette, the Opéra was used on three successive evenings to serve three different social pleasures, which became the ritual for all later marriage fêtes given there, often enough for the royal bastards. On the first night, by the use of remarkable machinery operated by ropes, the orchestra floor was raised to the level of the enormous stage and a gala dinner was served; on the second night, with the floor lowered, the Lully opera was performed; and on the third night, the floor having been hoisted up again, the theatre functioned as a ballroom for minuets. Closed after the Revolution and left to molder, the theatre was opened a few times by Louis Philippe in the nineteenth century to fête visiting minor royalty, and once by Napoleon III, for a dinner for Queen Victoria. After its recent renovation, it was reopened with the great Rameau divertissement "Les Indes Galantes," which was offered to Queen Elizabeth of England on her visit to Paris last spring. It seems worth noting that the troubles from which the Paris operas are now suffering to extinction began at the Versailles opera house last October, when the state sceneshifters sent from Paris first struck, causing to be cancelled the fashionable première of a new Cocteau ballet, "La Dame à la Licorne," since which no performance has even been attempted. In 1875, the Constitution of the Third Republic was proclaimed in the theatre, and no visitor to France's Fourth Republic should miss it, now that it has finally been made democratically visible.

Once again an intellectual, independent Left Wing—though anti-Communist—periodical has been suppressed by the police, on order of the French government, because it put into print certain seemingly incontestable unpleasant reports on exceptional phases of the conduct of the Algerian war. This week's *France-Observateur,* edited by Claude Bourdet (son of the noted author of social comedies), has been seized, in the fifth suppression in twenty months, for its publication of, and horrified comments on, the tortures, mostly by electricity, that were inflicted by French parachutist officers in Algiers on Henri Alleg, the French former editor of the *Alger Républicain,* which was politically against the Algerian war. The next day, Bourdet brought out an edition with the offending Alleg passages omitted but containing Bourdet's protest at the seizure and also at the loss of revenue—around twenty million francs—that the various seizures had caused. The Alleg torture case has become so widely discussed and so troubling an issue in France that Les Editions de Minuit, the little Resistance publishing group of the Nazi Occupation, recently published the whole book of Alleg's horrifying experiences, written in secret by him in prison and called "La Question"—the question being: Should the French torture political prisoners? From this book, which is still freely circulating, Bourdet's *Observateur* reprinted what it claims is nothing like the worst part of Alleg's gruesome story. An earlier issue of Jean-Jacques Servan-Schreiber's weekly *L'Express* was suppressed for giving similar information on the parachutists, who, as Alleg discovered, are the French Army's specially privileged, tough combat units. François Mauriac, the Catholic Academician, in his back-page *Express* column called "Bloc-Notes," has also written with Christian horror of the tortures. Nor has this French government action against the freedom of the press—a freedom basic in civilizing French concepts since the time of Voltaire—been confined to independent French papers only. Two January, 1958, numbers of our conformist *Saturday Evening Post* were temporarily impounded by the French police because they contained articles written for the *Post* by a former New York *Times* man who had easily gone behind the Algerian battle line to visit with the Front de Libération Nationale troops. He merely described their highly organized setup for supplies, weapons, food, and so on, but the French government apparently did not wish to have this presented to American or bilingual French readers here as having any reality, since Robert Lacoste, the French government's Minister in Algeria, has for months been saying that the war is "in its last quarter of an hour." International journalists here have also been scandalized that Miss Nora Beloff, Paris correspondent of the respected old British weekly *Observer,* was recently given, in private, a spirited, well-informed governmental calling down for her critical articles on the conduct of the Algerian war, the complaint against her having passed from the French Embassy in London to the British Embassy in Paris—all very polite and formal and menacing to the freedom of the responsible international press.

An understandable exception to this government censorship was made in the case of *Le Monde's* courageous publication in December of the extremely important report on French—and also Algerian—excesses in Algeria, as compiled on a tour of investigation by a distinguished French civilian Commission for the Safeguarding of Individual Rights and Liberties. This was indeed a public-spirited *Monde* service, since the paper printed this report, already much rumored about, without government authorization three months after it had been presented to the government, which had promised to publish it and then had done nothing. Nor did the government risk doing anything to the sternly reputable, influential *Monde.* Especially impressive in the text was a December, 1955, quotation—made long before the safeguarding commission had seemed necessary—by the then director of the French Sûreté Nationale, reporting on Algeria to his superior in Paris. He said he found it intolerable and embittering to human relations between the French and the natives that "French policemen can recall by their behavior the methods of the Gestapo," adding, "As a reserve officer, I cannot bear to see any French soldiers compared to the sinister S.S. of the Wehrmacht."

France-Observateur, in both issues of this week, reports that for the third consecutive time delegations from the British Labour Party have demonstrated on Saturday morning in front of the office of the French military attaché in London, to denounce "the parody of French justice" at the recent trial of the well-educated Algerian woman Mlle. Djamila Bouhired. Accused of complicity in terrorist bombings, she was legally condemned to death by a French court in Algiers, where—most illegally, it is claimed—her counsel was not permitted to plead for her. The British Labourites have been demanding that President Coty reprieve her from her approaching execution, and in Oslo the president of the Norwegian High Court of Justice, along with many judges, two prefects, one Army general, and a pleiades of professors, writers, poets, and so on, has demanded a revision of the Bouhired trial, if only for the sake of France.

Since the French Army's deplorable bombing of Sakiet, which was a real and great shock to most French—except the majority of the French deputies, to judge from their lack of bold, angered comment in Parliament—the war in Algeria has significantly increased in activity, on both the rebel and the French sides. Typical is one of this week's front-page communiqués in *Le Figaro,* announcing "very violent fighting and ambushes in Algérois and the Constantinois," with a hundred and sixteen rebels laid low, and forty-five French military men killed and eleven missing, which, *Figaro* adds, is "the worst fate of all, perhaps, considering the more savage natives' mutilating form of revenge against the Western white man who still defends the remnant of France's North African empire." Never in modern times, fortunately, have so few men on both sides been killed in a war, but this war has been costing the French the gigantic outlay of a billion francs per diem, thus practically bankrupting the state without the war's being won. The present so-called "good offices" of our Mr. Robert Murphy in trying to settle the differences between the French government and President Bourguiba of Tunisia seemed to be progressing usefully until the French Foreign Affairs Ministry announced that good offices from foreigners could not include stopping this forty-month-old persistent, envenomed, and perhaps final white and Arab war.

May 25

For the past fortnight, the French Fourth Republic has functioned in what has seemed a state of unrealities that have become facts, upon which it still survives, with difficulty. Wednesday evening, the pro-Gaullist newspaper *Paris-Presse* said, with irony, "To sum it all up, the situation is now clear. Parliament has confidence in Pflimlin, who has confidence in General Salan, who has confidence in General de Gaulle, who has no confidence in Parliament but is waiting for it to show confidence in him." *Paris-Presse* then undermined its concise gibe by adding that things here in France "are obviously less simple, less tragic, and less absurd" than that. Actually, the tragedy, the lack of logic to the point of farce and fantasy, and the intricate complexities have all been major, dominant, equal, and probably unavoidable since May 13th, just twelve days ago. That was when the contagious Algiers insurrection broke out against the Paris government, soon coming under the command of a military and civilian junta to which General Raoul Salan, chief of the

French forces in Algeria, gave his blessing—a troubling double performance in which the General is still starring, rather like a trick cavalryman aloft, riding with one foot on the back of each of two dangerous, plunging horses. The strain on parliamentarianism and the Fourth Republic was increased by General Charles de Gaulle's stately offer, repeated at his press conference last Monday, to take over power in France—an offer that, as the days go by, makes him seem the inevitable, unique, and perfectly non-Republican solution to the Fourth Republic's present political drama of paralyzation.

In the accumulated confusions, one thing alone now looks clear. For the first time in peace, Paris is not the political capital of France. Temporarily, at least, the colonial city of Algiers is the French political epicenter, is making the violent, dominating political news, is claiming to be the revolutionary leader "of the renewal of the French spirit," and has been making the active decisions, to which Paris is helplessly susceptible. Even if in the next few days, or possibly weeks, President Coty, equipped with all due legality squeezed from the Paris parliament, should offer de Gaulle the leadership of France, Algiers would still triumph, because its citizens—by the hundreds of thousands—lawlessly shouted for him first, crying, "De Gaulle to power! *Vive de Gaulle!*" This is a call for the savior that has not yet been massively heard here on the Paris streets.

The two weeks' growth and emerging aim of the Algerian uprising constitute the most peculiar colonial rebellion of modern times. The dissidents' goal—since it was a movement from the Right, drawn from Algeria's million French *colons,* or white settlers, who are the country's prosperous racial minority—was not to obtain independence from France but to attach themselves indissolubly to it, in order to make sure that they would not be abandoned by it through the unsuccessful ending of the three-and-a-half-year-old, ill-fated French war against the liberty-loving *fellagha* rebels, who are the activists among the nine million Moslem majority. In case you have not been able to unravel the tangled Algiers events since May 13th, it seems that on that afternoon there was a solemn, angry memorial service at the city's Monument for the Dead to honor three French Army soldiers killed by the Algerian nationalist forces in reprisal for the execution of native soldiers by the French. After this patriotic service and the chanting of the "Marseillaise," a gigantic riot occurred, in which thousands of the French citizenry, instead of assaulting the Moslem quarters, as usual when enraged, sacked the French Government Building, aided by anti-*fellagha* Moslems. According to the latest newspaper reports here, the riot was the opening battering ram of a well-laid plot by a group of hard-core *colons,* who had distributed tracts all morning to both the white and the Moslem populations. The tract urged an uprising against "the worn-out system," meaning the weak Paris governments of splintered political parties, and especially the Pflimlin government. Pflimlin, the tract said, was scheming to hand over Algeria to the *fellaghas* in a negotiated peace, stranding the *colons* in costly independence from France—a money-losing, undesired liberty indeed. There was also, perhaps, a less well-laid plot for May 13th among certain professional French Army officers, long bitter against what they have called the inefficient, ignorant, unpatriotic Paris politicians, who, they claim, engineered their defeat in Indo-China and, peace being more popular with most French voters, have been hamstringing the Algerian war. Friday, the independent Paris paper *Le Monde* printed a May 13th letter that the youngish Brigadier General Jacques Massu, of the crack, brutal 10th Paratroop Division, sent to his whole division, officers and men, which at one point said, classically, "The hour is grave." Then it assailed the Paris stabs in the Army's back, "aimed at dishonoring us and weakening our accomplishments," and asked that the letter be sent to the soldiers' families in France. "You, our parents, and our friends in Metropolitan France have an obligation to help us in denouncing the campaigns against the Algerian war by so-called intellectuals, lay or Christian, and those who have lost their national sentiment and now play the game of the foreigners, whether of the East or West," it said—meaning Russia and the French Communists, who call it "the dirty war," and the United States, with its anti-colonial and supposedly pro-Algerian foreign policy. That May afternoon, Massu, with his red beret, his Cyrano de Bergerac nose, his brown-and-green camouflage battle dress, and his jumping boots, was reportedly a noticeable stalwart in the frenetic Algiers crowds, which, as the excitement, violence, and rioting continued, became host to a new phenomenon—fraternization between *colons* and Moslems in a sudden brotherly unity that, after a hundred and thirty years of French occupation, is already being called "the miracle of Algeria" and that, to common cries of "Long live French Algeria!," is still going on all over the land.

To restore order among the delirious mobs, General Salan took command of the uprising, and just before midnight issued a communiqué saying, "Having the mission to protect you, I have provisionally taken in hand the destinies of Algeria." By May 15th, the General was himself shouting with the crowds for de Gaulle. Since then, Salan has been nightly addressing throngs of about fifteen thousand and invoking the installation of de Gaulle from a rostrum that now flies the tricolor of France barred with de Gaulle's Cross of Lorraine. Three nights ago, on Thursday, when the still wild and enthusiastic crowds shouted their newest slogan, "The Army to power!," he thanked them for *"cette bonne parole"* (for "these good words"), adding, "We shall all march up the Champs-Elysées together and be covered with flowers."

Two days ago, a newly enlarged Comité du Salut Public de l'Algérie et du Sahara held its first meeting in the Algiers Summer Palace, with a red carpet, an honor guard of spahis holding sabres aloft, and General Salan in the chair, surrounded by delegates from the Departments of Oran, the Sahara, and Constantine (some of them natives who had been flown in), and by about three times as many French *colons* as Moslems. General Massu and Mohammed Sid Cara, former Paris deputy from Oran, were elected joint presidents. The official report of the meeting declared that the Salut Public movement had done more work in three days than any Paris government had done in three years, and added that it was its firm resolution to set up in France "a public-safety government, headed by General de Gaulle, to demand a profound reform of the Republic's institutions."

Yesterday afternoon, according to a prepared speech by Pflimlin that, inexplicably, was first broadcast on the national French radio at two-fifteen this morning, when the French were asleep, the Prefecture of Ajaccio, in Corsica, was captured by force by Pascal Arrighi, Corsican Radical Socialist deputy to Parliament here, on his return from a visit to Algiers, an action in which he was aided by some local Ajaccio citizens and a hundred and fifty French Army paratroopers stationed there. Premier Pflimlin came almost out in the open in his speech when he said that "despite the culpable conduct of certain men," the fraternization between French Europeans and Moslems had "given hope for the birth of a new Algeria"; but that there was no such pregnant excuse for illegality in Corsica. "I have not the right to hide from you the fact that the same danger exists in

France," he went on. "It is my duty to warn all those French who are attached to their liberties as guaranteed by the Republic's laws that sedition-mongers here are attempting to drag us to the brink of civil war."

As seemed natural and heartening, Pierre Mendès-France made the major diehard liberal speeches recently in the Assembly. He not only denounced the Algerian generals but declared that the more important general, Charles de Gaulle, had "aggravated a dramatic situation" in his press-conference speech last week. Mendès said that during the war he had been a follower of de Gaulle, in which he took some pride, and had always thought that someday—Mendès himself being a critical pessimist in political affairs—de Gaulle would once more become the artisan of a national reconciliation, fatally due in France. "However," Mendès said, "sedition has just broken loose in Algiers. Civilians and military men took a weighty decision in invoking the name of de Gaulle. In the face of insolent sedition," Mendès said boldly to Pflimlin, "your strength is your legitimacy," as representing Republican law. "It is the Republic that entrusted these military chiefs with soldiers to command," he went on, "and it is time that these chiefs were recalled to a sense of their duty and honor. . . . General de Gaulle should resuscitate the emotion that he arouses in the people in order to strengthen the Republic, for which he now reserves his severities—after, admittedly, having restored it to life following the Liberation."

Half a million men of the French Army are in Algeria, and a handful are in Germany; Paris not only is without protection but, were it not for the military men who now operate as the official bridge across the Mediterranean, would be without communication with Algiers, its alarming godchild. Pflimlin is calling a special Assembly session tomorrow, to be followed by his bill for constitutional reforms, which comes up Tuesday. It may not pass, since, with desperate intelligence, it cuts the deputies' Parliament-sitting to only five months of talk, sets the limit of a government's life at two years, and prohibits the overthrow of any government unless a new Premier and new government are all ready to replace the old one. This constitutional reform has been talked of for twelve years, was cooked up in committee in four days, and is slated to be passed in one week—thus being twelve years one week and four days late in contemporary history. It is supposed that Pflimlin will be followed by General de Gaulle.

De Gaulle's enemies will automatically be the Communists. There are well-informed Parisians who think that Moscow, with its telescopic long view of the world, will not order French Communists into street riots against him. The Muscovite vision, they say, sees any republic as a rotting fruit that will sooner or later be soft enough to drop into their hands—a process that de Gaulle, by his personal flame, might only hasten. Militant Communist Party members declare, however, that if he comes to power they will fight against him in the streets—"our skin against his"—despite his having been a war leader whom the Communist Resistance greatly served and admired. If the Communists choose to fight, they will paralyze France by strikes. The Socialists, though they have hardened in the last few days against de Gaulle, will doubtless accept him, since the choice for them now is de Gaulle or a revived Front Populaire with the Communists, in which the minority Socialists would be eaten alive, as if by political cannibals. The hour is indeed grave for the Republic. Much depends on General de Gaulle.

To judge by his recent statements, he still has no sense of human relations and no personal ambitions—nothing but his sense of sacred destiny, which has seemed to be an intermittent one. He is, as a political poet recently said, like a stained-glass Chartres window—an awkward, colorful, large male figure seen against heavenly light, a glassy symbol of devotion that, because of its optical distance from human observers, prompts them to reverence. A giant of a Frenchman in outer and also in hidden ways, he has just been critically described by Jean-Paul Sartre as "the interminable man." Loving nothing but France, he is the Frenchman of sacrifice.

June 1

Never since the time of Mme. de Sévigné have so many high-styled, personal, historic letters about vitally important public events been written and rapidly sent off in France, to become the talk of the nation, as have been written here this past week. On Thursday afternoon, beginning promptly at three o'clock, the reading aloud of one of them by the Speaker of the Assembly, addressed to him and *MM. et Mmes. les Membres du Parlement* by M. René Coty, President of the French Republic, at a moment of national destiny, was rewarded by a scene of such fury, tumult, and party passions as will leave a deep scar on the annals of that institution. Speaker André Le Troquer, standing narrow-shouldered in his evening clothes—which are the Speaker's prescribed attire, night or day—and addressing the Assembly in his fast, experienced, smooth voice, first invited the tense deputies filling the hemicycle to rise to listen to the message from the highest official of France. As everyone now knows, the Coty letter was a despairing appeal to the Assembly to call upon General Charles de Gaulle to form a legal, exceptional government to save the crumbling Fourth Republic. "Now we are on the edge of civil war," read Le Troquer in his quick, skimming tones. "On one side and on the other, people seem to be preparing for fratricidal strife. What will become of our France?" Then he went on to this passage: "In the moment of danger for the country and the Republic, I have turned toward the most illustrious of Frenchmen, toward him who, in the darkest years of our history, was our leader for the reconquest of liberty . . . one whose incomparable moral authority would ensure the salvation of the country and the Republic. I ask General de Gaulle to confer with the head of the State. I shall then, in my soul and conscience—" At a sign from the Communist leader Jacques Duclos, on the front bench of his party's seats, the entire Communist group of about a hundred and forty men and women deputies sat down, in one obedient, interrupting action, on hearing the phrase "soul and conscience," as if the utterance of such words in connection with de Gaulle and power had sent them a little underground, in company with many neighboring Socialists, who seated themselves in belligerent accord. At such gross manners, a deputy from the Center contemptuously shrilled *"Goujat!"* ("Scum!") at the Left.

The Coty letter came to an end after his promise, *"en mon âme et conscience"* to resign from the Republic's Presidency if his attempt failed. Instantly, pandemonium broke loose. The Communists beat their desks with their hands, bellowing insults and threats, shouting "Fascism shall not pass!," and shaking their clenched fists at the other deputies in what looked like genuine human hate and destructive rage. At a further signal from Duclos, they suddenly jumped to their feet and burst into "La Marseillaise," while the other deputies (except for those sympathizing Socialists) remained seated, as a mark of indignation, but gesticulated and twisted around to scream epithets at the Left. Somebody in the topmost visitors' gallery let fall a cascade of tracts onto the heads of the Christian M.R.P.s sitting directly beneath. The Communists had already started another song, "Le Chant du Départ"—the celebrated Revolutionary hymn written to honor the fifth anniversary of the fall of the Bastille and sung even under Napoleon's early rule. At its most famous line—*"Tremblez, ennemis de la France!"*—the Communists lifted their arms, like a choir trained in elocutionary gestures of hatred, to point to the deputies on the Right, and at the end of the song they rapidly walked out of the Assembly in a body, the session having been declared at an end by Speaker Le Troquer. Then the members of the Pflimlin government walked out; the Poujadists, of the extreme Right, marched out, also chorusing the "Marseillaise"; and the rest of the hemicycle emptied itself quickly. The whole scene, from the Coty letter on through to the last note of the last song, had taken exactly fifteen minutes by the big Assembly clock.

De Gaulle's most important and characteristic letter of the week was dated Wednesday, an hour before midnight, and was written in answer to one of harsh criticism sent to him by the Socialist former President, old Vincent Auriol. The General said, in part, that "the Algerian events were provoked, as you well know, by the chronic impotence of French governments," and that in his desire to serve the country he had struck "determined opposition" in the Assembly. "As I could not consent to receive power from any other source but the people, or at least from their representatives," he wrote, "I fear that we are moving toward anarchy and civil war. In this case, those who, moved by a sectarianism incomprehensible to me, will have prevented me once more from saving the Republic will bear a heavy responsibility." Then emerged his great, tragic final phrase, in the style of the seventeenth-century French classic dramatist Corneille. "As for me," the General wrote, "there will be nothing more until I die, except to dwell in my sorrow."

If this vitally influential trio of letters, each written on its own altitude of character and of patriotic love, best clarify the week's events when read chronologically backward, the rest of what was said, secretly or overtly, during the week was entirely confusing. To put the main and opening fact simply, the Socialist Party members not only were aroused by Auriol's letter to intransigence against de Gaulle's coming to power but were also partly in rebellion against their own Party chief, Guy Mollet. In early 1956, Mollet was elected Premier on his Party's regular platform of pacifism and anti-colonialism, but once he had been hit by tomatoes and dung thrown by Algiers *colons* on his first visit there in February of that year, he speeded up the Algerian war against the native rebels with a military severity that even Rightist deputies had not dared demand or hope for. Furthermore, Mollet this week favored de Gaulle's candidacy after a secret, pleasant exchange of letters with the General, which he hid from his followers until Pflimlin let the cat out of the bag in Parliament. However, behind his back, his Socialist Party members had organized their mutiny against de Gaulle without telling him, either.

The week's events soon turned into the regular farces, dramas, and nocturnal Assembly sittings of an exceptionally desperate French political crisis, plus fatiguing late nights of conference for the elderly important men like Coty and Auriol, and even de Gaulle, who made four three-hour motor trips between his country house in Colombey-les-Deux-Eglises and Paris in his old black Citroën—a total of more than a thousand miles, partly done in the dead of night at great speed—accompanied by motorcycle outriders and pursued by exhausted reporters and photographers. One day, there were three tiring sessions in the Assembly, which meant nine flights of stairs to be climbed by those of us faithful listeners who sat in the modest topmost visitors' gallery. At the Tuesday session, Communist Duclos made the most able, if hypocritical, speech for the defense of the Republic, and old Socialist former Premier Ramadier mumbled in his gray beard about the perennial piety of Socialist actions. On Thursday, when Paris became alarmed at the long delays and, through the English papers and the B.B.C. (never more popular here, what with the censoring of the French press and radio), learned of the Algiers threats of a showdown and of Algiers planes being poised for flight to France, one of the earlier days of inaction was called "Chloroform Day" by the Communists, since the de Gaulle push in Paris seemed anesthetized. At one point during the week, Mollet and a fellow-Socialist named Deixonne actually took a private plane, on de Gaulle's summons, to fly out to Colombey-les-Deux-Eglises and talk with the General, whom Deixonne, obviously surprised, later described as "nothing like Louis XIV." Motoring back to the airstrip, they were held up on the road by a herd of cows.

On Thursday, the false rumor was bruited about—two days too soon—that Coty had accepted the resignation of Pflimlin's government, which indicated (also too soon, it turned out) that through last-minute Socialist aid de Gaulle would have a comfortable parliamentary voting majority for his Assembly investiture and could come to power. From about ten that night until the early morning, well-to-do young French, in good and sometimes spectacular cars, drove in an endless carrousel up and down the Champ-Elysées and around the Place de la Concorde and other districts, tooting their horns (forbidden for several years), either because they enjoyed tooting them again or because they were in favor of de Gaulle. In any case, they always produced the same three short, two long squawks, which supposedly could be decoded to mean *"De Gaule au pouvoir"*—a gay but silly and provocative performance. The next night, when they were giving a sonorous encore, Communist militants armed with shovels—an odd nocturnal weapon for the Champs—smashed some cars and faces. Result: seven injured, no one killed, many of both factions arrested.

The most significant demonstration of mere people—as distinguished from Assembly activities—against de Gaulle was Wednesday's gigantic late-afternoon march of probably a quarter of a million men, women, and students, mostly bearing signs that said only *"Vive la République!,"* from the Place de la Nation to the Place de la République. It was organized by a suddenly collected group of resistants to de Gaulle, called the Comité d'Action et de Défense Républicaine; the Communists were not invited to take part, but they forced their way in. Beneath the majestic carved columns on the Avenue du Trône, set up in the time of Louis XVI, tracts were distributed by professors and their students, the demands being dignified and restrained. They said, in part, "General de Gaulle's last declaration leaves no doubt; it is a defiance to the workers and the people of all France. The question is not now one of either government or a constitution but of our most elemental liberties. Freedom of speech, of public assembly, of unions and their right to strike is today menaced by military power." The demonstrators marched by professions, by trades, by tragic experiences they had shared in the war. One old ex-schoolmaster carried a homemade wooden sign with bits of bread tied to its top, illustrating the ancient French cry "I am defending my bread." French Negroes from the lower African territories marched in affiliation with all groups. The most attractive section was the women lawyers, mostly nice-looking, who got heavy applause from the sidewalks. There were workers from the Renault automobile factory, some still in their dungarees and pushing the bicycles they had been about to ride home. Among the professors marched Francis Perrin, France's most distinguished physicist. Among the intellectuals were Jean-Paul Sartre and Simone de Beauvoir (frequently cheered) and the movie and stage star Gérard Philipe (whom the sidewalk girls screamed for). The most dramatic group was the former inmates of German concentration camps in their blue-and-white-striped uniforms, like pajamas for a last fatal sleep. As the marchers turned into the Boulevard Voltaire toward the Place de la République, the Communist elements filtered in with their slogans and signs, and the strictly republican character of the march changed. There were signs and shouts of "Down with de Gaulle!," "Down with Fascism!," "Put de Gaulle in a museum!," and, referring to his long neck, "Put the giraffe in the zoo!" Wisely, the police had been given orders not to interfere. It was the biggest, most orderly assembly of street demonstrators in Paris since the war.

Last night, it seemed almost—but not quite—sure that de Gaulle would go before the Assembly today to be invested. His appearance was announced for ten o'clock this morning, and then for eleven, and then it was said that he might not go at all if he failed of the majority of four hundred that he demanded in advance. The ceremony was finally set for three this afternoon. The afternoon started out as a fine, dulcet sunny one for this tense national event, so anxiously desired by millions of French, so angrily impeded by other millions. By two this afternoon, it took at least a press card even to stand on the quais across from the Assembly. Armed police in vans had assembled without sounding their sirens and were lined in silence everywhere around the Parliament. Behind it, in the Place du Palais-Bourbon, front rooms in the Hôtel de Bourgogne et Montana had been taken by a few French ladies and some photographers who wanted to see the General emerge from his car. He had said earlier this week that he had "a phobia about the Assembly," but he was persuaded that his presentation for investiture might be counted illegal—and also that he might be thought lacking in courage—if he failed to mount the tribune to state his elementary program. For the ceremony, he was dressed in gray, and he looked waxen and weary as he made his brief, succinct, intelligent speech. He said, in part, that "the government I shall form—granted your confidence—will immediately set to work on constitutional reform." Such a reform, if he gets it, will doubtless give him the powers of an American President, which he has always insisted were needed by a Premier of France. He defined the three basic principles of republicanism in France: One, that universal suffrage is the source of all power; two, that the legislative and executive elements must be separate; and, three, that the government must be responsible to Parliament—which (as in Pflimlin's earlier proposed reform) he said he would send on a six-month vacation, to leave him free to govern without its constant gabble and interference.

During de Gaulle's short speech, the Communists sat completely quiet. There had been a top-drawer rumor that Moscow, through its Paris Ambassador, would tell the Communist Party to do just that and then make a few forceful street demonstrations, though not forceful enough to have their own men killed. Shortly after three, the General finished his speech and retired from the chamber to await the Assembly's vote. In the Tuileries, shock troops of the Compagnie Républicaine de Sécurité, in blue uniforms, gently cleared the gardens of people, starting from the Place de la Concorde. The guards, with their carbines on their shoulders, pushed the few lingering adults along toward the Louvre exit in a leisurely manner and corralled the last straggling children who had paused for a final kick at their footballs. By six o'clock, the Tuileries was empty beneath its trees, even of the armed guards. The voting for or against de Gaulle was still going on. A thunderstorm suddenly broke the good weather. The sirens of police cars were audible along the Seine, but there was no sign of any anti-de Gaulle manifestants—perhaps too dampened to push on to their goal, the French Parliament. Finally, at eight o'clock, the news came that the General had been elected by a pitifully small majority of 329 to 224. The evening ended in a vigorous dark storm.

June 8

In the almost unbelievable public changes that have taken place during the past two weeks, General de Gaulle, who is victoriously responsible for them, has himself been changed in the process into M. de Gaulle. Properly speaking, that is. In becoming Président du Conseil (in the Third Republic more classically called Premier), de Gaulle as governing head of the Fourth Republic automatically outranks himself as General, and must take the superior title of M. le Président. The result, so far, is that he is simply referred to as both, eighteen years of his fame as *le Général* being too strongly associated with his militant, patriotic personality for the public and the newspapers to drop it easily.

The newly formed Cabinet of M. de Gaulle, as everybody expected from his contemptuous opinion of most parliamentarians during these last, fatal years of their disservice to the Republic, furnishes—in part, at any rate—a welcome, important novelty for France. A fourth of his Ministers are men who have done many things but at least have never been deputies. His Minister of Foreign Affairs, M. Maurice Couve de Murville, was, until last week, French Ambassador in Bonn, and before that in Washington, and thus knows American and German policies inside out—highly useful experience for this unusual new French government. He is, furthermore, a strong United Nations and NATO man, and this may quiet Washington's fears that de Gaulle will be intransigently nationalist. Another de Gaulle Minister with Quai d'Orsay ties is M. B. Cornut-Gentille, former Ambassador to the Argentine, who as yet has no fixed portfolio. The new government's Minister of Defense, M. Pierre Guillaumat, has been a topflight functionary in national bureaus dealing with atomic energy, mines, and, above all, oil, the last being of greatest importance to France since the recent Sahara oil discoveries. De Gaulle's forty-six-year-old *chef de Cabinet,* M. Georges Pompidou, after a precociously successful bureaucratic career, has for two years been director of the Rothschild Bank.

The Minister who is inevitably arousing the most public interest and curiosity is André Malraux, so-called Minister of Information. The information services are indeed under him, but actually he is something rarer, called Minister Delegated to the Presidency of the

Council and meaning de Gaulle's Special Minister. Informally and characteristically, he has just referred to himself, according to one Paris newspaper, as Minister of Urgent Affairs. The most renowned literary figure in France today, an *homme engagé* since youth, a legendary romantic adventurer in the early days of the Chinese revolution and in the Spanish civil war, a Resistance hero, the Goncourt Prize author of the great revolutionary novel "Man's Fate," and the most brilliant intimate admirer of de Gaulle, whom he served briefly as Information Minister in the post-Liberation provisional government and later as a phenomenal rally speaker for his ill-fated R.P.F. political party, Malraux is, to boot, the most compulsive, stimulating conversationalist in French intellectual circles, offering the widest panorama of political and social ideas. It is therefore not surprising that, since he is always extremely quotable, fragments of some of his conversations apropos of the "urgent affairs" of de Gaulle—and of the nation—over the past two weeks should, through the reports of friends, be frequently heard in Paris.

Three days ago, Malraux is known to have said to a friend, speaking of de Gaulle's aim, "It is capital that this revolution of French institutions—because, let's face it, 'revolution' is the correct word—should be accomplished without the shedding of a drop of blood. You understand, if the price of what he is trying to do were to be even a thousand dead in Algeria, for the General that would be too much." When the friend asked "And what if this bloodless revolution fails?," Malraux reportedly replied, "Then the forces that only he can hold in check will set to work within fifteen days." It will not be possible, in that event, to escape civil war, he went on to say, and France will be lost. It will be the way it was in Spain; civil war will not be localized but will break out regionally, and, weakened as the country already is, it will be ravelled away—and North Africa, too—until nothing is left. It will last three years; there will be a million dead. There will be no more France as a great nation; it will be finished.

The outstanding general criticisms of de Gaulle's Cabinet were against his choice of the recent Christian M.R.P. Premier Pierre Pflimlin and of the former Socialist Premier Guy Mollet, but in Algiers there is acute and dangerous resentment against all the Cabinet members who represent what the Algiers insurgents call *Le Système,* meaning all the familiar political faces from a now discredited Parliament—all, of course, except Deputy Jacques Soustelle, himself a member of the Algerian junta. This week, Malraux reportedly said, speaking of de Gaulle's struggle with the Assembly, "Though the parliamentary spirit is fairly foreign to me, I must say that Ministres d'Etat Pflimlin and Mollet showed from the first day an admirable civic spirit." It is known that they and President Coty, who has been described as "even more Gaullist than Malraux," having all been profoundly alarmed by the French state's collapse of authority and deeply convinced that only de Gaulle could save Algeria, and perhaps the country, from anarchy and bloodshed, acted as good-will bargainers between the Assembly and the General. Day by day, they ascertained what the recalcitrant deputies demanded. When the Assembly agreed to grant the General special powers to govern by decree, it ceded them for France but balked at letting him have them for Algeria, where he considered them an absolute necessity. When Mollet carried to him the message to that effect, the General is reported to have said *"Bon,"* stating that if he was invested Sunday, he was supposed to go to Algeria within a few days, and if a vital decision arose there, was he supposed to hurry back to Paris and convoke the deputies so as to meet the emergency? *"Eh bien,"* he added, it would not be he who would come to Paris to convoke them, it would be the parachutists. For he would be dead.

According to men around Malraux, the General with difficulty determined on patience in dealing with Parliament's reckless, talkative dawdling, even though he considered that every day's delay helped the junta forces in Algeria that were organizing against him, which, if the Assembly would hasten to face up to the inevitable, he could check. His patience with the Assembly, like his determination to avoid anyone's shedding a drop of blood in his name, was, in his mind, to serve as proof, which the whole country could see, that those who called him a dictator, or a Fascist, pursuing power, had either lied or erred.

As for the General's aims in Algeria, Malraux has reportedly said that "even if this recent new Franco-Moslem fraternity was in the beginning more or less organized"—even *truqué,* some French declared—"it is clearly apparent now that it is an extraordinarily popular and powerful movement, with a historic sweep." To make a new Algeria, two elements are already available, he thinks—the French Army and the Arabs. Astonishingly, according to him, the Arabs fraternize with the soldiers, who are all young, and call them "the Frenchmen from France," in distinction to the *colons,* or

Algerian French settlers. And the soldiers fraternize with the Arabs as a way of dissociating themselves from the *ancien régime* in Algeria, with its fifty or more dominant wealthy families, who practically own whole *départements,* such as the winegrowing Département de Bône, and are now considered a diminishing part of the Algerian political picture. For an important new social pattern is supposedly being formed. What de Gaulle will try if peace is established, Malraux is said to have declared, is to launch a really big fraternal operation between the Arabs and the young soldiers.

As for the French Communists, who will be de Gaulle's most tenacious local enemies, it is said to be Malraux's opinion that the Gaullists think they can considerably enfeeble the Communists' already rather weak attacks (the Party never having recovered its full popularity in Paris since the Hungarian suppression) by taking a series of real measures, or doing things, as he expressed it, the way they're done by the Anglo-Saxons, who believe more in realities than in words—and who are right, too. In a nutshell, by fighting Communist ideology with facts. Apparently, de Gaulle does not think it useful to fight against Communism in the name of any ideology, even a denunciation of capitalism, since mere capitalism is no longer the acute question. It is, instead, necessary to fight by setting forth objectives for the working class that have hitherto not been possible under French capitalism. Furthermore, on June 18th, the eighteenth anniversary of de Gaulle's historic broadcast from London to France, in which, to however few listeners, he announced that France would fight on, and thus founded the Resistance, Malraux, evidently counting on bringing to bear what is in theory a brilliant psychological coup, plans to have the General hold a parade not on the Champs-Elysées, where the official parades of Paris are traditionally held, for the bourgeoisie and what is left of the old carriage trade, but in the Place de la République, for the populace that arrives by Métro; that is, he will hold the parade for the working class in its neighborhood stronghold, thus perhaps changing the geography of public feeling and partisanship.

Who or what is really responsible for the horrifying crisis of fallen republican authority in France? To this question, which one hears on all sides in Paris, Malraux has a curious answer, according to friends' reports. In large part, he says, the accumulated collapse has been due to the taste of the French deputies for mere talk. When the General appeared before the Assembly, he spoke simple truths and was impressive: Certain things are going to have to be done to get the country out of the trouble it is in; if you don't want to do them, we won't get out of the trouble; in that case, I'll simply go back home to my village of Colombey. In the Assembly, at the end, even Duclos, who is ordinarily a precise speaker, spoke, says Malraux, for a half hour without saying anything. His speech had no reality; it was an abstract discourse on Communism which was not Communist, against a dictator who was not a dictator. It is this garrulousness that has made a farce of the republican parliamentary form.

Just before the General flew to Algiers, he is said to have remarked to Malraux that he was an old man, to which Malraux replied that he did not seem to be one. The General reportedly then said he would have energy till the day he died, like Clemenceau, but he had the feeling within him that he was old. He said he had done what he could—once. The result of what he was doing now would be only to give a certain glow to France. If he succeeded this time—which at the moment appeared possible—and if for French youth his *patrie* could before he died become a reality, then, despite all, he would have brought about the real Liberation. General de Gaulle was, Malraux is said to have declared, in vivid description of that moment, in a state of both hope and despair that was not far from true grandeur.

September 24

During General Charles de Gaulle's weekend of speechmaking in four key provincial cities—Strasbourg, Rennes, Bordeaux, and Lille (his birthplace)—on behalf of his new constitution, which is to be voted on in a referendum this Sunday, he realistically told the immense crowds, in his toneless but trumpetlike voice, that his aim was "for France to keep her place in the hard world of today" with modernized governing institutions. As one innovation, his constitution could prevent Parliament from making French governments fall like autumn leaves (but several times a year). In his speeches, he mentioned France's need for greatness but did not refer, as he used to, to her old habits of glory. With further realism, he begged his listeners for "a massive number of yes votes, a crushing majority of yes votes, which would be a proof of faith in him who now speaks to you—faith that he needs," he added loudly, in that stately, confidential manner with which he refers to himself in the third person, as if he were his own public witness. For this referendum, each of the possibly twenty-eight million voters (who have had to register specially for the exceptional balloting event) is being sent a four-page copy of the new constitution, so that its ninety-two clauses can be perused. Though the referendum has only acceptance or rejection of the constitution as its immediate goal, the mere monosyllables of "yes" and "no" will, in an inferential way, have a three-way spread. The yes votes will mean that de Gaulle should be kept in power as Premier until November, when, it would seem logical, he himself can be voted on as a new kind of powerful, executive President of the Republic. The Sunday yes votes will also be an affirmation for a legitimatized Fifth Republic, which de Gaulle represents in advance—as, indeed, he seems to represent everything salvational now in sight.

A sharp-witted old Paris charwoman recently said to a foreign acquaintance, "*Bien sûr,* my husband and I will vote for de Gaulle, but without enthusiasm." A youngish Paris businessman voiced the same prevalent feeling by saying, "I shall vote yes and think no." A millionaire industrialist more candidly said, "Certainly we don't want him, but we will vote for him. Better vote for him than for a revolution." Three weeks ago, Parisians already were saying that everybody would vote for de Gaulle, though nobody was a Gaullist, and that he would get a landslide eighty per cent of the votes. This figure has now been more temperately estimated at around sixty-five per cent. Not even his bitterest political enemies think that he can lose. The French will vote for him in masses because in this catastrophically troubled year for France he is the unique, untarnished great figure on their horizon to turn toward; they will vote for him by the attraction of historical gravity, with none of the magic delirium of adulation that gilded his majestic popularity after the Liberation, when France was at his feet.

De Gaulle's weekend provincial crowds were the first to be enthusiastic. The main reason the French have mostly lacked open enthusiasm for him seems to be that he worryingly reminds them of themselves and of their ancestors, and of those special figures in history to whom their ancestors rallied, with increasingly diminishing returns—Napoleon Bonaparte and then Louis Napoleon, who, at his own request, was transformed by the French citizens' vote from a President into the shape of an Emperor. This Second Empire recollection has inspired de Gaulle's political opponents—who consider his republicanism suspect and his constitution's strengthening of the President's powers dangerously Bonapartesque—to nickname him the Imperial President, the Prince President, and the Monarchical President. Last Sunday in Lille, on a wall of the house where his father, a professor of philosophy and literature, lived, and where, sixty-seven years ago, Charles de Gaulle was born, the local Communists derisively scrawled "Charles XI born here," putting him in direct succession to Charles X, France's last Bourbon king, who believed in absolute monarchy. Something from all these characters to whom the French at times have mistakenly rallied has been rubbed off, by a kind of cruel decalcomania, onto the electoral picture of de Gaulle, including a supposed likeness to General Boulanger on his black horse and, obviously, to de Gaulle's arch-enemy, the anti-republican, autocratic Marshal Pétain. Today, in the confused light of all these historic warnings, pressures, and hopes, mixed with many Frenchmen's latent shame for their veneration of Vichy and the instinctive French suspicion, ever since the Dreyfus case, of all high-up Army men in political life, General de Gaulle alarms people's imaginations as their prospective Fifth Republic political leader, just as his unquestioned probity, lofty intelligence, and Gothic character rouse their faith in him as a man and as the only figure—or so they believe—who can morally hold France together.

The greatest deterrent to de Gaulle is that he was brought back onto today's scene by the bulk and cream of the French Army in their Algiers *coup d'état* which successfully invoked his name as its borrowed banner. "At the present moment," comments a brilliant British correspondent, writing with a detachment that the French now lack, "liberals and diehards in France are waiting to see which way the General will turn after the referendum. He has left the diehards an almost free hand in Algeria for four months. He could have done little else without revealing the limitations of his own power. . . . He has nonetheless kept alive the hope that the policy of Algiers is not his." The danger of civil war that de Gaulle faced four months ago, with the prospect of insurrectional French paratroopers dropping on undefended France, is the same civil-war danger, with the same *paras,* that is being conjured up now—plus Communists fighting against them from behind Paris street barricades if de Gaulle loses and if, as is his habit at dead-end moments in his personal history, he simply goes back home to Colombey-les-Deux-

Eglises and leaves France flat. What the liberals here profoundly fear, on the other hand, as do the French politicians, is that if de Gaulle wins with too massive a vote he will be a presidential prisoner of this extreme Right, with no sizable liberal or Left Wing parliamentary opposition to help maintain his liberty. Pierre Mendés-France denounces the civil-war threat as pro-de Gaulle political blackmail. But (or so many French believe) he says this without taking account of a very real possibility of civil war—the civil war that might come if he, with all his remarkable political intelligence, but with his special unpopularity, were now faced with the task of holding Catholic France together.

Obviously de Gaulle's most potent enemy and the leader of the vote against him is the Communist Party, with its morning paper *Humanité's* scurrilous cartoons of him and its latest hypocritical chicanery—its slogan that the Communists are "the defenders of the Republic," which they have successfully helped to scuttle. The rich industrialists and the big businessmen, who will vote yes, are against de Gaulle as a man who wants to change things, which alarms them, so *Le Figaro,* which represents the conservative bourgeoisie, has been straddling the fence. Only today, four days before the referendum, it came out with a melodramatic editorial urging a vote of yes—less for de Gaulle than against the Front Populaire and a possible civil war. The General's most effective opponents are the left-of-center intelligent liberals, whose daily spokesman is *Le Monde,* the best-written analytical newspaper in France, and the sharp, courageous weekly *L'Express*. Little *Combat* is the only hysterically pro-de Gaulle paper.

Naturally, "yes or no" has become the new phrase of France—in talk, in night-club songs, in advertising. Thousands of the big and little organizations that are basic to Frenchmen's instinctive centralization of life and to their appetite for controversy in groups are now busy declaring how they will vote on the referendum. Newspapers print their decisions in adjoining columns, headed "Yes" and "No," as a means of showing which way the election wind is blowing. Thus, you can read in your daily papers that the Primary Schoolteachers Union and the University Committee for Republican Defense will both vote no. But the National Association of Parents of Public-School Pupils and the National Committee of the Middle Class will vote yes. The Friendly Society of Former Deported Jews of

France will vote no, along with the National Federation of Former Deportees, Internees, Resistants, and Patriots and the Communist Friendly Society of Widows, Orphans, Descendants, and Victims of War. The Comte de Paris, official pretender to the throne of France but personally a Socialist, has published his monthly public communication as a stately little essay entitled "Oui." There is even a Civic Action Front Against Abstention, which hands out free publicity for window displays from its office on a boat, tied to the *quai* below the Pont de l'Alma. Guy Mollet's Socialist Party decided in caucus, with a certain ambivalence, to vote yes, except for a little splinter group that pulled loose so it could vote no. With similar ambiguities, the Radical Socialist caucus also registered a majority decision to vote yes, though its most important member, Mendès-France, will, naturally, vote no. Higher up, five Cardinals of France have declared that the absence in the constitution of "all reference to God, obviously painful to a Catholic," is no reason to neglect the imperious "duty to vote" in these moments of supreme decision for the nation. The constitution's second article merely says, "France is a republic, indivisible, secular, democratic, and social," without adding that it is Christian. One prelate had earlier said that good Catholics should not vote on such a godless document. The Cardinals' decision, with its urgent tone, implies that the invaluable Catholic vote will be yes.

To date, the most searing denunciation of de Gaulle, and of the events leading up to and now surrounding his present position, is that of Jean-Paul Sartre, called "The Constitution of Contempt," of which the first installment was printed recently in *L'Express*. Nothing so far has equalled its talented vitriol, nor is anything likely to, unless Sartre surpasses himself in the promised second article. He opens with sarcastic *brio* by saying, "To begin with, who suggested this plebiscite? Nobody. It has been imposed on the sovereign nation. . . . Our referendum enjoys the doubtful charm of being impromptu. . . . They began by trampling our old institutions underfoot; now they propose this elderly frippery, a royal charter"—his synonym for, and opinion of, the General's new constitution. "The voter lost in the no man's land that separates the dead republic from the future monarchy has to make up his mind all alone; it is all or nothing, 'all' being King Charles XI and 'nothing' being the return to the Fourth Republic, which nobody wants any more." Speaking with disdain of the General's characteristic pride, Sartre adds, with mordant humor, "I do not believe in God, but if in this plebiscite I had the duty of choosing between Him and the present incumbent, I would vote for God; He is more modest." After tracing the events of the French Army's May *coup d'état* against the Paris government as foundation for de Gaulle's present position, Sartre says that nothing and nobody "can make us forget that General de Gaulle was carried to power by the colonels of Algiers," and continues, "We are promised a return to calm, to discipline, and to tradition, provided that we give our votes to the rioters of Algiers. Let us not fool ourselves; all the plebiscites in the world cannot prevent a *coup de force* from being, and from remaining, a form of disorder. The barrel will always smell of herring; the Gaullist regime will always smell—to the end of its days and in all that it does—of the arbitrary violence from which it sprang." He ends with a call to vote no and a final warning to the General, saying, "On one point we are in accord with you: the Fourth Republic is dead, and we have no notion of resuscitating it. But it is not for you to make the Fifth Republic. That is for the French people themselves, in their full and entire sovereignty."

If further proof were needed of the weighty importance and extraordinary inclusiveness of Sunday's vote, it should be added that this simple yes or no, when in the hands of the millions of dark-skinned inhabitants of France's thousands of square miles of deep African territories, will mean for them a vote of "Yes, we wish to remain attached to the new de Gaulle French commonwealth" or "No, we prefer our independence from France." In a really astounding application of logic to the premise of self-determination of peoples, de Gaulle has given all these territories the right to have independence merely by asking for it—provided they can afford such freedom, for they will, if free, automatically cut themselves off from all French financial aid and administrative guidance. One old Dahoman tribesman is reported by a traveller to have said sagely, "The white General does not understand. We do not wish freedom as a donation, we wish to wrest it from him." It is amazing that the French and their politicians have accepted almost without demur this fabulous invention of de Gaulle's for avoiding further ruinous and unsuccessful wars against men wanting independence. True, Deputy Jacques Isorni, the diehard nationalist lawyer who defended Pétain in his trial, wrote an indignant piece in *Le Monde* complaining about the General's giving away pieces of France as if it were his personal property. And Pierre Poujade, chief of the shopkeepers' political party that bears his name, expressed himself in simpler commercial terms, saying, "He is disposing of the empire as though at a bargain sale." Nigeria and French Guinea have already stated in advance that they will vote no to any suggestion that they "lack responsibility to run their own affairs," meaning that they choose independence.

In Algeria, of course, a heavy yes vote will be cast by the white French *colons* in support of Algeria's integration with France. But Algeria, being considered part of metropolitan France, is the one overseas region where a no vote does not constitute a vote for independence. So the rebels' National Liberation Front, which has been fighting for nearly four years for the country's independence, is aiming at preventing the Moslem population from voting at all, to judge by a report of their terrifying orders for voting day in the rural sections. These say, "It is forbidden to go to the market place. In forest regions, the population is to go to the woods. The roads will be mined and road traps set up, and corpses [presumably of those who headed for the market-place polls instead of the woods] will be placed where they can be seen. Ambushes will be laid for vehicles sent for voters far from the towns. Through our control system in the villages, we can know who has disobeyed these instructions. They will be punished one way or another."

The acts of Algerian terrorists, increasing in violence as their desperate protest against the coming election, are like a fanatical fringe of the Algerian war that has been carried all over France and into all sections of Paris. Two Sundays ago, a stabbed Algerian lay dying in a pool of blood on the sidewalk before Lipp's *brasserie* in St.-Germain-des-Prés, with a crazy old flower vender shuffling in the gore and offering his faded bouquets for sale, as if it were all a scene from a Surrealist picture. Minister Jacques Soustelle was almost murdered at the Arc de Triomphe, and now a dynamite bomb has just been found in a ladies' *lavabo* at the top of the Tour Eiffel.

What lies behind de Gaulle's revolutionary change in French republican history is the problem of the Algerian war, which brought the Fourth Republic to its knees, as the French know and say. On the Algerian problem, the General has so far said not a word—not even the word "integration." What the French are waiting to see, they declare, is what he must inevitably say and how soon he will say it, to save France.

October 1

The towering majority of referendum yes votes given this past weekend to General Charles de Gaulle—almost eighty per cent in France and a fraction more than ninety-six per cent in Algeria—is so impressive that it has changed France's relations not only with herself but with the rest of the world. Certainly the Western democratic peoples now eye France with real hope for the first time since November, 1945, when the General initially took power for what turned out to be merely three months. And the time before that—the final opportunity in a regular Republican France was Sepember, 1939, or nearly twenty years ago, at the start of the war, which is a long, long time for doubts. Even the General, with his oracular, intellectual truthfulness, says that his new constitution is not perfect but "corresponds to the necessities." *Le Figaro* has just remarked, "Even an excellent constitution, which no country has, cannot suffice to reconcile different political points of view." Deputy Pierre Poujade, with his vulgar knack of sometimes hitting the nail on the head, has just said, "With the help of God, who is certainly also a Gaulliste, [the General] is now going to try to reconcile the irreconcilable and give satisfaction to Christians and Marxists, to liberals and collectivists—in short, to all sections of his new majority, or to nobody."

What all these power dissensions produce is the scandalous, bigamous accumulation of France's sixteen political Parliamentary parties, which the French continually deplore and vote for. Though often bastard fractionizations, they are based on six political conceptions that represent profound, legitimate differences of contemporary Gallic opinion on man's relation with government, and are the real, separate, and specific essences of French political thinking—essences and conceptions so ineradicably opposed that already many French are saying that they must deadlock, if not strangle, the Fifth Republic, even under the exceptional authority of de Gaulle. As the public and the politicians admit, only panic fright and helplessness to do anything else temporarily led four of these divergent political forces to unite long enough to vote the Fifth Republic in over the weekend—that quartet being the old anti-clerical republican force, the political spine that divides the French body politic toward the Left and the Right; the pro-clerical Catholic force; the forces of the

Right; and the Socialists, a mildly Leftish, respectable brand here. The two recalcitrant forces that refused to take cover and, as a consequence, voted no were the intellectual Left liberals and, of course, the undeviating Communists. It would seem that the French—the most logical, literary-minded nation, made up of the most individualistic citizens, whose brains and fertile political imaginations invented the eighteenth-century French Revolution, thus influencing and redating all the Western world—are today, in the twentieth century, bringing their own republicanism to impotence not only through breeding too many theories but through the overuse of logic. To all this, their inherited medieval taste for schism has added the final weakening touches. In June, *Le Monde* announced, with its customary authority, that "the Fourth Republic died of widening splits, which separated the boundaries of the political parties from the real frontiers of public opinion." As far back as three years ago, M. Edgar Faure, then Premier of France for the second time and one of the nation's shrewdest younger political brains, said publicly, "I want to speak my conviction, resulting from personal experience, that in the present [multiple-party] condition of the regime, government of France is not scientifically possible. The problem is not, as is often said, one of governmental instability. It is one of governmental impossibility. The governments here are unstable because they are impossible." This is the fractioned French political state that very likely even de Gaulle's constitution and new electoral laws cannot greatly stick together. In any event, that will be his second difficult task. He is now working on his first and even more important difficult task in Algeria, which voted its almost unbelievable ninety-six-per-cent-plus for integration with France in the face of the bankrupting Algerian war for independence.

The new Paris UNESCO headquarters, in the curious Y-shaped building on the Place de Fontenoy, behind the Tour Eiffel, has long been counted on to show, when completed, the most spectacular manifestation yet of modern art as decoration, produced by a galaxy of the most famed international artists. Well, practically the whole set-up, including the creations of the Spaniards Picasso and Miró, was opened last week, and the public, the professional art critics, and even modern-art lovers do not like it, by a loud majority. In Picasso's eighty-square-metre fresco—slapped onto plasterboard panels and put up crooked (so the panel figures don't jibe perfectly)—there is one white disc-faced female with triangular breasts, of his war-period style; one recumbent, foreshortened male figure (who, through economy of outline, is shaped like a pretzel) baking himself in the sun; and one half-female, seen only from her ballooning hips on up—the implication being that the rest of her is hidden underwater, as she is apparently wading. In the center is a characteristic postwar-Picasso bouquet of what looks like four white skeleton arms and fingers on black. What is sure is that when the afternoon sun brightens the fresco's plain blue and green-blue lower panels, they look as if they had been painted slap-dash by Picasso's children, with brush strokes going every which way. Miró's contribution of a short, thick ceramic-brick wall that separates nothing from nothing in an empty garden space is decorated with red spots, ticktacktoe stars, black rings, and a blue moon. "Why so much art that looks like playthings?" an irritated Paris critic asked, including in his disgruntlement the American Alexander Calder's gigantic two-ton black mobile near the Avenue Suffren, which passing children do indeed pause to admire. Yet if you get a beeline perspective on the building and the Tour Eiffel beyond, a kind of artistic metal aerial brotherhood is established between the two—the single connection that UNESCO has with the neighborhood. The English sculptor Henry Moore's garden figure, with holes, and the French sculptor Jean Arp's abstraction are not yet placed—perhaps luckily. The Italian Afro's painting outside the seventh-floor restaurant was called "a discolored daub"; the Chilean Matta's canvas in the bar was said to look like "unidentifiable machines."

The uncontestedly admired aesthetic contribution is the large Japanese garden, planned by Isamu Noguchi, on the Avenue de Ségur side, with its upright green and gray stones worn to contemplative shapes of beauty by time at work on nature, and with its dwarf conifers—both stones and trees having been presented to UNESCO by the Japanese government. There are panels of running water atop walls, and an arched stone bridge leading nowhere except into the imagination. In the apogee of criticism of the whole UNESCO building and its contents, one art critic finally said sharply, "This new aggrandizement of UNESCO demanded the destruction of eighty-three trees on the Place de Fontenoy. The future will decide the worth of this functional architecture, denuded of aesthetics and sensitivity." This seems to bring us back to the old-fashioned criterion that only God can make a tree.

October 15

For any connoisseurs of curious brief periods of French history, this present one should indeed be memorable—an important, fascinating hiatus to be living through, to watch and listen to, in which the larger, shaping events are only now beginning to show and be heard from. Since the referendum, the French public, usually volatile during exceptional political scenes, has seemed to be struck almost dumb by a mixture of relief, shock, and hope. The main definite new fact is that the Fifth Republic now legally exists, without a cheer's having hailed its birth, since it was merely born twice on paper, in two different editions. The first birth was on Sunday a week ago, when General de Gaulle's constitution was published, for anyone to buy, in the state's inexpensive brochure of records, *Le Journal Officiel,* its appearance there having the automatic effect of making it the supreme law of the land. On Monday afternoon, it had its more elegant second birth in the Empire dining room of the Ministry of Justice, in the Place Vendôme. There, as tradition demands, the seal of state and the legalizing Ministerial signatures—though Premier de Gaulle himself was too busy to come and sign at that moment—were affixed to the single, historical, deluxe copy of his constitution, to be saved for posterity in a glass case along with the dozen or more French constitutions created since 1791. This special copy, bound in gold-embossed scarlet leather, was splendidly printed by the *Journal Officiel* printers in Jaugeon type on vellum. Neither on the red cover nor in the text does this newest constitution indicate that it is for a Fifth Republic. However, the succinct few words spoken after the signing by the Minister of Justice dated it tragically and accurately. He said that the constitution's aim was "to create a great communal association, above races and religions . . . in a century . . . of violence and acute racial struggle"—an unmistakable reference to today's Arab Moslem and French white-Catholic Algerian war. He narrowed down his time sense even more when he spoke, as though in warning, about the glass box full of opposed French constitutions: "The ordeals and quarrels of France lie written there. We must admire the permanence of the nation despite its lacerations. We must admire the permanence of the state despite the unparalleled shocks to its power.

But let us make clear that nobody should, as a continuing practice, stake all on a miracle, because one has always to pay high for a miracle, and it is possible that one day a miracle could fail to happen."

De Gaulle is now the fully functioning miracle in France, and it is merely the defunct and abused Fourth Republic that is paying history's high price for him. Veiled by his own legends but unobscured by any power politics of Parliament (closed until January), he temporarily holds all the viable power in his own hands, free of the melees of deputies still back in their constituencies but busy preparing themselves for recrudescence in November's reduced elections—all of which makes this probably the last time to see fairly clearly what kind of wonder-working qualities he has brought to France's shabby political scene. According to many Parisians, what he has really brought is the result of his long life—his repeated tragic experiences, as a high patriot, with French history, when he seemed to be the only one who knew which way French history was going and could not stop it. To this past is added the psychology he has drawn from his classic military education. In his final chance now to fix up modern France for survival—or so he must hope—his precise contributions to his government from this double past have been his elderly sense of dynamic realism, his impatient haste to get important things started before it was too late, and his infallible instinct for strategy. Notably in Algerian affairs, he has realistically been pushing France, in his constant loud-voiced speeches, toward humane reforms and generosities, which he thinks can alone accomplish her reconstruction and salvation in North Africa. But the problem of ending the Algerian war, which has bled France's finances white, he has treated only strategically—that is, with his typical complete silence, no one having maneuvered a single word out of him, in public, on that subject. For his strategy in important affairs is to ignore what is not ripe enough to talk about. What he will say to this last weekend's humbled but tremendously important offer from M. Ferhat Abbas, the leader of the Algerian government-in-exile, he is still being strategically silent about. But Parisians are discussing Ferhat Abbas' proposal with relief and hopefulness. It is the rebels' first offer to negotiate a cease-fire without demanding, as a precondition of negotiation, recognition of their people's independence, for which, to date, they have been fighting three years and fifty weeks—a *cessez-le-feu,* Ferhat Abbas said, "to bring to an end the indescribable sufferings that this war has already caused." The N.L.F. rebels consist of—at the outside—perhaps a hundred thousand Algerian fighting men, whom the French Army's four hundred thousand soldiers have not been able to defeat. If the Algerian war ceases, it will be one more bitter end to a little war fought in minor history.

Algerian affairs being once again more important in France than French affairs, the other major strategic silence de Gaulle has been maintaining is on the word *"intégration,"* which he has refused to pronounce in his speeches here in metropolitan France as well as in Algeria, where it was considered an insulting omission. Its particular meaning, especially important to them down there, is that sandy North African Algeria is actually a geographical extension and physical part of France. *L'intégration,* the main idea that the Algerian rebels have been fighting, is, of course, the belief of Algeria's million French *colons,* who believe in white-man supremacy, and it is also the key dogma of the Algiers colonels' junta. Their chief, General Raoul Salan, recently avowed that General de Gaulle had said to him, "Algeria is and will always remain French soil." General Salan further observed, in order to make clear his belief in the complete geographical naturalness of integration, "The Mediterranean traverses France the way the Seine traverses Paris." One of the Paris papers immediately published a wonderful cartoon map of Paris, showing it traversed by the Mediterranean instead of the Seine, with the Place Pigalle located in the north of France, the Opéra in France's middle, the Place de la Concorde down in North Africa, the Gare de Lyon somewhere east of Suez, and Montparnasse deep in Africa's jungles.

Ever since May 13th, when the junta took power, liberal Gaullists here and all over France have been waiting for their General to speak up, to dissociate himself from what seemed neo-Fascist French military men, who, by shouting his name, brought him to power. They have been waiting for him to prove he was not the Algiers colonels' property as a dictator—or, indeed, a dictator at all—and, by denying the colonels their intimacy with him and by throwing them from power, to raise himself aloft alone, the elderly liberal leader in an old general's uniform. In his new instructions to Algiers, he put them in their place—beyond the pale. Tuesday morning, when these sensational communications were read on the breakfast radio news program and appeared on the front pages of the daily papers, lots of French began telephoning each other about *"ces bonnes nouvelles."* Many French who had voted for him but were against him in fear of his becoming their autocrat seemed almost glad to be proved wrong. Among liberals and pro-Gaullists, the General's liberation of himself—doubtless finally made possible on the strength of his showing in the referendum—has aroused admiration and the first complete satisfaction as to his republicanism since his equivocal entry into power. In his brief but astonishing personal letter to General Salan, he wrote that the government felt that the best interests of Algeria required the November Algerian elections for Parliament to take place "in conditions of liberty and absolute sincerity," and added, "Candidates representing all the political tendencies [meaning the rebel N.L.F. candidates, as well as any others] should be allowed to solicit the votes of the electorate competitively and on an equal footing. You will, in consequence, conform to the directions here enclosed. *Croyez, mon cher général, à mes sentiments cordiaux,* Charles de Gaulle"—a very cool ending.

In the enclosed instructions to the Army leaders came the retribution for the Algiers French military men in the peremptory order to put an end to their powerful regime: "Furthermore, the moment has come when members of the Army must cease taking part in any organization of a political character, whatever the reasons that may have furnished exceptional motives for joining. . . . Nothing from now on can justify their belonging to such organizations. A report shall be made to me on the measures taken to execute these instructions."

In the referendum, de Gaulle won the most smashing victory over the Communist Party ever registered and admitted by the Party since it became the biggest voting bloc in France. The Central Committee has finally published in *L'Humanité* a remarkably frank analysis of why the Communists lost a million and a half of their voters ("and we did not even feel it coming"), who were supposed to vote no against the General but voted yes instead. "One voter out of five did not listen to us," the report said bravely. "This is the first time since the Liberation that such a phenomenon has occurred." Among the reasons listed was "the real desire of the French people for a change," which is putting it mildly. Another was the Algerian terrorist activities in France. Still another reason that the French Party cited as perhaps influential was, after all, "the vestiges of anti-Communism after the events in Hungary." Then came what sounded like an almost confessional explanation: "To millions of French, humiliated in their pride by the servility of previous governments, de Gaulle seemed the guarantee of national grandeur and independence . . . the negotiator who would make peace. There reigned great illusions about de Gaulle in our working class." It appears that a million and a half of them had faith in him.

Seen historically, de Gaulle now seems to constitute the sole dynasty of modern France; for the third time, he has succeeded himself, after a due interval, as the country's leader. He has the ability to be automatically reborn to power through France's dire needs or tragedies—as in last May, as in August of 1944, as in June of 1940. Over the dozen years since early 1946, when he angrily walked out of his last leadership, he has gradually turned into the portrait of an entirely different man. The rural seasons at Colombey-les-Deux-Eglises have added weight that minimizes his impressive height; an operation on his studious eyes has cost them their former haughty glance; in the hollow toning down of his voice, it no longer leaps sometimes to falsetto, like a trumpet played out of tune, as it used to do. With his waxen features and lengthy nose, his face is treated by French caricaturists to look like a melted candle. But when he speaks, that candle still burns fiercely. In this, his third functioning as France's savior, he seems to the public to have taken over France's identity—he at the age of sixty-seven, and she, owing to her recent batterings, nothing like as young as she appeared after the Liberation. In his speeches during these past months, he has perhaps more often than usual referred to himself in the third person, as if absent-mindedly throwing aside the disguise of being anything except the voice of France herself. The purity of his patriotism, this mystic feeling of his that France is his mission, was what the eighty per cent of the people voted yes for, not being sure of much else at the moment. His exalted, romantic passion for France, which he used to treat almost like his private concern—fervently defended, like any man's personal passions, against all intruders or sharers—he now presents differently. It is something that he actually seems to share with, not just define to, the French people today. An austere, superior rationalist and eccentric, descended from long generations of privileged, educated minds, he never used to be at ease with gregarious democracy or with the idea of sharing his elevated sentiments with forty million other citizens. A reasonable misanthrope, he did not love people by the million. Now, in this peculiar hiatus that separates the two republics, it is the millions who have really helped him save his *belle France* and start her on her way again. After his lonely, thoughtful years at Colombey, humanity appears to have become his broadened concern. In his recent speech in Constantine, which the French feel marked the new de Gaulle as humanitarian and statesman, he offered a magnanimous five-year plan for Algeria's uneducated, impoverished citizens—provided, of course, that the French can afford to pay for it. Above all, where he was once obsessed by the idea of reëstablishing the glory of France, he is now merely aiming at its unity.

In this last fortnight, he has gone a long way toward achieving it. He has set up a true voting democracy for Algerians, for the first time in the two races' common experience. By his now famous brief communications to General Salan and his military, he has overthrown the Army's power politics in Algiers and the supporting *colons'* white-supremacy ideal, which together in May sacrificed the Fourth Republic, rather than face the long-due Algerian peace negotiations that de Gaulle is now free to consider. By this republicanism, he has restored Paris as the political capital of France, has freed himself from strategic ambiguities, has thus aroused a unifying approbation from the major political parties and the men who were still waiting to see and believe, and has himself suddenly become the greatly popular leader of France. In the first two weeks of his Fifth Republic, General de Gaulle has taken his biggest active place to date in the procession of history.

October 29

The sex of the automobile in France—where sex is of the highest importance, especially verbally—has been officially settled. As of the last fortnight, the automobile here is a he instead of a she. This change of gender was proposed a couple of weeks ago by Academician Jean Cocteau to the Académie Française, and accepted, without public explanation of what must have been the valid etymological reasons, in one of its scholarly lexicographic séances, such as led to its great first dictionary, in 1694, and now aim, through slow studiousness, at bringing one out on twentieth-century French, too—possibly around the year 1999, to judge by the rate at which it is creeping through the alphabet. Whatever the Académie decides, however, immediately becomes correct usage. Right after the First World War, the French automobile was already tending to be slightly androgynous, with the 1923 edition of the big, two-volume Larousse dictionary defining it as a feminine noun but adding, "Some make it masculine." These few, if still alive, would have been right in style again talking about *un bel auto* at the Forty-fifth Salon de l'Automobile, which has just closed at the Grand Palais, and where, for two weeks, ordinary Frenchmen tried to pick out *une belle auto* as usual, she having become completely female and increasingly desirable over those intervening years. As one gazes today around Paris—until a few seasons ago a city of open vistas and spacious beauty, now a night-and-day necropolis of inert parked cars on every historic square and nearly every street—it is tragic to ponder the fact that the Salon's natural function is to stimulate the sale of the thousand French cars a day that are being turned off the assembly lines and that absolutely must be sold for the economic good of France and the expanding aesthetic ruination of her loveliest old cities and towns.

The winter literary season is now under way. As the publishers recently announced, in what sounds like a warning, one out of every three novels that will be printed in it is by a woman. So far, only one remarkable new volume by either sex has come off the presses—"Mémoires d'une Jeune Fille Rangée," by Mme. Simone de Beauvoir, which reads like a novel but is an autobiography, based on her private diary, of the conformist first twenty-one years of her life, in her polite Parisian prison of social and Christian obedience to her parents, from which only her precocious young brain and senses at first escaped, via the window of her mind, to soar with grave juvenile questions about God, love, and death. This was the primary compartment of her life—from her birth, on January 9, 1908, among the white-lacquered furnishings of her parents' bedchamber in their superior apartment on the old Boulevard Raspail, to her first meetings with Jean-Paul Sartre in the Sorbonne student cafés twenty-one years later, when her adult life abruptly began. One reason her autobiography is as emotional as a novel is that its basic theme is an individual's struggle for physical and intellectual liberty—the double imprisonment due to her being a well-brought-up upper-class young French female, of a "good but obscure name," with its ennobling particle. It was still that period of the French *culte de la famille* in which nice young girls were reared in the salon, in innocence and stupidity. Her family was halfway between the bourgeoisie and the aristocracy, toward which her lawyer father, as long as his money lasted, leaned by pretension and preference. Thus, what imprisoned Simone and her less fractious younger sister, Hélène, was really the leftover nineteenth-century ideas that ruled the upper-class French until long after the First World War, when the twentieth century finally overcame them. In certain fascinating pages of Mme. de Beauvoir's book, her childhood seems as far back as Marcel Proust's, though not so rich, and never unhealthy. She had high-spirited vitality, was violent, and was prone to what she calls *extrémisme*. Her father, who dominated her by his charm, despised the Republic, liberalism, Zola, and Anatole France, but read aloud to her Victor Hugo poems and Rostand, and also de Gobineau, that pre-Fascist French authority on the inequality of the human races. An anti-Semite, M. de Beauvoir was as convinced of the guilt of Dreyfus as her mother was of the existence of God. It was the disequilibrium between her father's worldly male skepticism and her mother's female piety that early drove the daughter to search for the proofs of ideas and that "largely explains why I became an intellectual." Indeed, at the age of thirteen, leaning one night out of the window of her grandfather's château in the Limousin, and already in love with love and nature, she painlessly lost her faith in God. Even then disillusioned by the lies, hypocrisy, and folly of adults and by the horrors of the war, she reports, "It was easier for me to think of the world without a creator than of a creator guilty of the world's contradictions"—certainly adult philosophical reasoning for the age of puberty. The void left by religion she filled with literature, reading, among other works, George Eliot's "The Mill on the Floss," in English, because it also dealt with a case of conscience, and wading through parts of her grandfather's library, including a seven-volume history of the Bourbon Restoration, which turned her into a democrat. At the age of fifteen, she knew that she wanted to be a celebrated writer.

Now that she has become one, it is her professional writer's talent that makes her autobiography read like a novel, through her elaborate portraits of her family and friends, who seem to become sad minor characters as the fiction of their lives trails along. The *belle-époque* summers at her paternal grandfather's modern château, with peacocks on the lawn and four horses hitched to *le break* for family drives, ended with his death. Her mother's father, a pious banker from Verdun, lost his fortune through crazy investments. Zaza, the girl for whom she had had a passionate friendship since their first schooldays together, died, mad of brain fever, when her bourgeois family opposed her marriage to the man she loved—another female victim of lack of freedom. Cousin Jacques, whom Mlle. Simone had loved and wept over for years, though with no desire for marriage—with her instinct for liberty and feminism, she scorned the position accorded to the French wife—at least enriched her life before eventually ruining his own and dying a penniless sot. That is, he gave her the new, modern France in books that, in her lonely family life, she had missed—Gide, Montherlant, Claudel, Alain-Fournier, Barrès, Mauriac, Cocteau, Jacob, and even Dostoevski, all of which her father loathed at a glance. M. de Beauvoir had long since lost his practice as a lawyer, vaguely going into insurance instead, and as the family finances slid downhill, he warned his daughters that they could expect no dowries and therefore could not marry but would have to earn their livelihoods, after which he regarded them as already *déclassées,* with Mlle. Simone headed for the Sorbonne to become a teacher, and Mlle. Hélène headed for art school. Studying ten hours a day a pell-mell mélange of Greek, Latin, English, Italian, mathematics, philosophy, and logic, and by sheer brilliance cramming four years of university work into three, Mme. de Beauvoir still had the energy for fits of *extrémisme* in Montparnasse night clubs—Le Jockey, Stryx, La Jungle—with mixed drinks, jazz, wild antics, and the first tastes of liberty, still hidden from her family's ears and eyes. In her last year at the Sorbonne, the three student ornaments in philosophy were Paul Nizan, André Herbaud, and Jean-Paul Sartre—a haughty, hermetic trio, followers of the famous Professor Alain, of the Lycée Henri-Quatre. The course of her friendship with Herbaud changed her life, for when he failed in his final exams and disappeared—and she passed—she came under Sartre's wing. He was, she reports, a man who thought every minute of his waking time, knew women as well as he knew books, had no intention of marrying, and had planned his life for travel and writing, for exploring the world and truth. Again as in a novel, this made the unmistakable happy ending to her youthful years of struggle, romanticism, and isolation, and to her longing to be equal, appreciated, and free. This first volume of her memoirs, which she will surely continue, is at once the changing history of modern social and intellectual France and the personal, preliminary history of a brainy woman who has helped change it.

General de Gaulle's impressive offer last week to the rebel Algerian Army of a *paix des braves*—a peace for brave fighters—united around him literally every shade of French political opinion, all filled with high hope that at last this dangerous, bloody, bankrupting racial and political small war was coming to an end. The rebels' unexpected sharp refusal—after the previous offer of a cease-fire by Ferhat Abbas, chief of the Provisional Government of the Algerian Republic, or shadow government-in-exile—killed the hope. It also constituted Premier de Gaulle's first setback in his miraculously successful opening month of Fifth Republic leadership. The double disappointment has been acute.

The war, which in French official semantics is called "the pacification," will enter its fourth year on Saturday.

November 13

Owing to the out-and-out novelties and the unparalleled confusions facing the French in their Parliamentary elections ten days from now, on November 23rd, the Paris political reporters can write only in a state of more or less complete muddle. There will now be seventy-nine fewer deputies for metropolitan France than there were under the Fourth Republic—merely four hundred and sixty-five seats, which de Gaulle hopes will make for a more manageable Chamber. For these seats—and a five-year term of office—there are 2,784 candidates now preparing for the scramble, from half a dozen major parties and about a dozen minor parties or splinter combines. In de Gaulle's arbitrary new slicing up of France into four hundred and sixty-five voting districts, each entitled to elect exactly one candidate, the winner will be the one with the majority on November 23rd, or with the plurality in a runoff election one week later. With an average of twelve candidates now running for every seat, it looks as if nobody could obtain a majority in such a panting crowd. By his single-constituency voting system, de Gaulle, of course, hopes to break up the feudality of the big parties' localized power and to favor the election of individual, responsible deputies, instead of the party-machine hacks or the intransigent party devotees, who, between them, brought the helpless Fourth Republic to its knees.

Certainly Premier de Gaulle, in a late-October press conference, warned the public and whoever was aiming to compose this new Parliament of what was facing them in the Fifth Republic. "After a certain number of months' suspension, the Parliamentary institution will reappear," he declared. (It is now thought that its reappearance will be in a special, brief session in January.) "But [Parliament] will no longer be omnipotent. . . . The future Assemblies will have precise limits and powerful brakes. . . . If it should unhappily come to pass that tomorrow's Parliament does not wish to accommodate itself to the role assigned it, there is no doubt but that the Republic will be thrown into a new crisis, from which no one can know how it might emerge—except that, in any case, the Parliamentary institution would be swept aside for a long time."

As further warning and advice to all the political parties that are trying to sail into power in the new Parliament on his coattails and under his name—their own party nomenclatures under the Fourth Republic being now so weakened in appeal as to need a timely refresher—de Gaulle said, "Everybody understands that I neither desire nor am able to lend myself in a direct manner to this competition. The mission that the country has confided to me excludes my taking part. Therefore I shall not favor anybody, not even those who have offered me friendly devotion through all the vicissitudes. *Bien entendu,* I shall not disapprove of those political groups that publicize their adherence to what de Gaulle has done," he interjected, in one of his most historic third-person references to himself. "But impartiality obliges me to insist that my name, even in adjectival form, be not used as a title for any group or any candidate." This prohibition has added the final confusion to the election campaign. This will be as strange an election as any French Republic has ever known.

The obscurities surrounding the elections in Algeria (in many ways far more important for France than those in France itself), where nearly no candidates for deputy turned up until just before the lists were closed, have produced in Paris almost a paralysis of interpretation and hope. In a belated rush, almost two hundred candidates finally appeared—either Europeans of the May 13th persuasion or Moslem Right Wingers, both in favor of Algeria's remaining French—to run for the sixty-seven seats in the Paris Parliament. Parisians can only suppose that the rebel Moslems' fear of identifying themselves publicly as candidates for the French Chamber, which might bring death by having their throats slit in reprisal, shows to what extent the ninety-six-per-cent yes vote that the native population gave to de Gaulle last month was false, owing to the vote's being steamrollered by the French Army there. It may be that, in having already offered them "a peace for brave men" and now Parliamentary rights—both refused—de Gaulle, remarkably experienced as he is in modern French history, has been too civilized to be immediately useful.

November 26

What the French people voted in favor of last Sunday was a single issue—the miracle-making quality of General Charles de Gaulle. He had his sixty-eighth birthday on Saturday, and from his Fifth Republic, which held its first Parliamentary elections the next day, he received the duly expected political gift of a strong swing to the Right—much more of it than he probably wanted.

Considering the peculiar confusions and circumstances of Sunday's election—only about four-fifths as many seats to run for as before; old political boundary lines broken down; new parties set up, with multiple hedgehopping in between; old faces searching to be rechosen; new, unidentifiable faces pushing forward to squeeze in; all the candidates whipped by more than the customary ambition and by a feeling that a new, last-ditch legislative period in France was taking shape, in which they had to participate or perish—considering all these factors, the results of the election have been illuminating, yet rather comic. A turnout of about seventy-five per cent of the nation's voters managed to elect exactly one deputy in Paris and only thirty-nine deputies in all France, since an absolute majority of the votes was necessary in order to win. So four hundred and twenty-six seats will have to be disputed again in the second-round runoff this Sunday, when a mere plurality will suffice.

It must be understood that this election, in which twenty and a half million votes were cast, really had no issue but de Gaulle. It was an election held utterly and entirely in terms of him, of this one Frenchman who was not even running for office. During the campaign, people turned up literally in twos or threes in the villages to hear the candidates speak; even in the cities, the bigger crowds cared little what was said, nobody really wanting to know one more professional politician's opinions on the vital topics for France today—such as the quarrel with England over the common market and with the French Minister of Finance for shortening the Sorbonne's educational budget—let alone wanting to hear any politician's notions on how to end the Algerian war. What the voters wanted to know was how the candidates stood on de Gaulle—whether they would back him, and whether they were running for an annual privilege of a very few months in Parliament on the truthful understanding that Parliament had failed to know how to govern and that only de Gaulle, by fiat and miracles, could now be trusted to direct the destinies and problems of France.

The only major party to campaign on a platform that decried de Gaulle and his constitution, the Communists got a million and a half fewer votes, all over France, than they did in the 1956 Fourth Republic elections. Their vote last Sunday was not quite nineteen per cent, which, by a tiny margin, still gives them the biggest cut of the nation's ballot. The Poujadists, also anti-de Gaulle, were practically extinguished on Sunday; they received less than one per cent of the nation's vote, enough maybe to elect half a deputy. Even ex-Premier Pierre Mendès-France, the single great political figure who, ever since last May, had been conducting an intellectual counter-offensive against the General, was defeated spectacularly on Sunday in his normal fief of Louviers, in the Eure, held by him since 1932. (Oddly enough—or so it is said—de Gaulle had hoped that he would be elected, so there would be one excellent intelligence in the parliamentary opposition.)

The winning mixture was the Union pour la Nouvelle République, or U.N.R.—initials that it might be well to learn now, since the party won close to eighteen per cent of the national vote and will figure predominantly in the coming Parliament, although nobody yet knows exactly how.

The new Chamber will add up to an absolute Rightist majority and a possible perplexity for de Gaulle in solving France's most vital unsolved problem, which is the Moslem rebellion for Algerian independence, just now entered into its fifth year. Much of the French Right—and, precisely, the U.N.R. leadership—stands for what is called the *intégration* policy in Algeria, which regards the country as a province of France and part of France's soil, without autonomy. The General has made it known that he favors a more moderate evolution, which could develop internal self-government for Algeria within the new French community.

Paris has just emerged from a five-week deluge of thirty-five concerts by fifteen nations, called Les Semaines Musicales de Paris and created by the UNESCO-sponsored International Music Council to honor the opening of the controversial new modernist UNESCO building behind the Tour Eiffel. The music lover's only problem was to figure out which dozen or more concerts he had the taste, the time, and the ticket money for. The city has probably never known such a daily variety of star players, orchestras, and conductors, of foreign musicians and foreign music, or such combinations of all of these. On one Salle Pleyel program, Ravi Shankar played the sitar, Shinichi Yuize played the Japanese koto, and then Menuhin and Oïstrakh played their Strads in Bach's D-Minor Concerto for Two Violins, but did not hit it off well together. The other outstanding disappointment was Stravinsky's new opus "Thréni," inspired by the Biblical prophet Jeremiah, which Stravinsky conducted. It was like a stony labyrinth, through which the listener wandered, pathless, without echoes, without pleasure. Before the performance, Stravinsky confided to an American musician that it was *"très ennuyeux et très savant"*—"very boring and very scholarly." True. At the Salle Gaveau concert by the excellent small orchestra that Queen Elizabeth of Belgium founded for its leader, Franz André—it was vainly hoped that Her Majesty would attend—the virtuosities of the violinist Arthur Grumiaux were a revelation in a Mozart concerto. Von Karajan, with the Berlin Philharmonic, packed the vast Salle de Chaillot, where he had played during the Nazi Occupation as junior replacement for Wilhelm Furtwängler, who refused to conduct in Paris because he had loved the city. However, the great old Hans Knappertsbusch, conducting the Vienna Philharmonic in Brahms' Third, displayed a discretion, coming from complete authority, that united his poetry and the instruments' sounds so that they all became one with Brahms himself. There were concerts by the Frankfurter Singakademie; a miners' chorus from Ostricourt; the Juilliard Quartet; an organ recital at St.-Sulpice by Marcel Dupré; the Czech Philharmonic; at least one piece of modern music included in each concert, so there was enough Webern and Berg; and one concert featuring "micro-interval music on sixteenth tones"—absolute bedlam.

The American Leonard Bernstein was the comet among all the stars. Heard of as a prodigy but not yet heard as a conductor, he accomplished the impossible at the second of his three concerts here when he played and directed the Lamoureux Orchestra in four piano concertos—a Bach "Brandenburg," a Mozart, a Ravel, and Gershwin's "Rhapsody in Blue." The appreciation that Bernstein aroused among music lovers and critics here was astonishing. Even Clarendon, the most trenchant of Paris critics, said, in *Figaro,* "Bernstein triumphed with an incredible ease over the multiple difficulties presented. To my mind, this session offered, like a mountain, two slopes. With the Bach and Mozart we were elevated to the summit of precision and poetry. With the Ravel we lost a little altitude, but with the Gershwin we were stabilized on a reassuring level. *Ne persiflons pas*—a success of this quality goes far beyond being a merely *sportive* performance. I regard Leonard Bernstein as a prodigiously gifted musician." At Bernstein's third concert, his deep and instinctive reading of Mahler's Second—without a score, and conducting an orchestra, vocal soloists, and a choir he had barely met—gave his listeners the impression of having heard an extraordinary local event in musicianship and pleasure. To his compatriots here, it oddly seemed that Bernstein, by his youthful generosity of gifts, momentarily restored among Parisians the popularity of certain American qualities.

December 11

In the last of the string of sabbatical elections here, on Sunday, December 21st, General de Gaulle will unquestionably be elected President of France—the personal, apostolic solution he has proposed ever since 1946 for bringing a strong single-headed government to this country. Even for the Fifth Republic, it seems really fantastic that only one candidate will be running against him—the Communist Party's token sacrificial goat, M. Georges Marrane, Red mayor of the suburb of Ivry, who will thus assure an election almost, but not quite, as limited as would be run off by the Communists in Soviet Russia, an ironic fact unappreciated by the French Muscovites here.

The new Parliament will not start functioning legislatively until the last Saturday in April, a good long time from now—an interval that its members, the French citizens think, can well use in figuring out how to act and survive better than their predecessors. However, this Parliament enjoyed its first temporary sitting on Tuesday, to elect its Speaker. The visitors' galleries were jammed to view this quiet, confusing, historic scene of gathering deputies (including one Algerian chieftain in gold-and-white turban and flowing robes). The moment awaited by all was the entrance of Premier de Gaulle, white-faced, dressed in dark gray. The deputies rose with a burst of applause as he took his seat on the front-row government bench and then mounted the little staircase to the tribune to shake hands with the eighty-two-year-old deputy and priest Canon Kir, the Assembly's senior deputy, who was to make the address of welcome. Only after the General sat down again, the cynosure of all eyes, did the Parliament cease politely standing. In his address, the old canon—cassock mussed, hair awry, spectacles perched crooked on his nose, an eccentric figure long known in the Chamber for his humor and ineptitudes—said that he "piously saluted the memory of those no longer with us" (meaning all the deputies who had failed of reëlection); that all candidates, like himself, had suffered from the public's violent criticism during the campaign; and that this new Parliament would have to fight such public opinion, since it could rapidly lead to "a dictatorship, which the great majority of the French do not desire." Somewhere, from some deputy, there arose a faint whistle.

For the spectator, this was a nebulous experience, this first sitting of a new Parliament of predominantly new men, of whom the nearly two hundred U.N.R. deputies "hardly know each other by sight," as *Le Monde* commented, with its customary trenchancy. The paper added, "Among them all, they have only one tie—Gaullism, a state of mind much more than a doctrine. As for a program, there is none. There is nothing but an unlimited adherence to the acts and even the intentions" of one man—de Gaulle himself.

December 29

In General-President de Gaulle's nationally televised Sunday-night broadcast concerning his government's dramatic financial measures, he looked to be much more vigorous than on the recent afternoon when he appeared at the temporary session of Parliament. His voice, the characteristically awkward, sudden gestures of his hands—as if discarding something—his whole mien, and his mental and physical presentation were energetic, authoritative, and certain. There were perhaps one or two occasions when he seemed to ad-lib, as if to make his lofty statements clearer on rapid second thought, in a speech that lasted precisely fifteen minutes and that he had written himself, as is his custom, and then committed to memory—this last an amazing feat of brain and concentration for a man of his age. It must be understood that if President Eisenhower's state of health was at one moment of sufficient emotional concern to the United States to affect the stock market, de Gaulle's condition of body and mind is of far vaster importance to France, because, on his own terms, he is literally all that the country now possesses for the state to rest on. In the opening portion of his Sunday speech, he said that the repeated recent elections and the referendum had confirmed him in his task, and added, "As the guide of France and chief of the republican state, I shall exercise the supreme power." He spoke pungently, being gifted with French literary talent and with a style formed by rich and extensive classical reading. Among other notable sentences was his statement that if the country "has charged me with leading it, this is because it wishes to go forward, certainly not toward what is facile but toward effort and renovation." He also said, with a touch of his dry humor, that of late France had been vacillating "between drama and mediocrity." Of the devaluation, he said, "Furthermore, for the old French franc, so often mutilated by our vicissitudes, our wish is that it regain the substance conformable to the respect due to it." The key of his speech was that his financial measures could base the French nation on what he called, in a near pun, *"vérité et sévérité"*—thus reuniting for his listeners these two grave, intimately related words, which "alone can permit France to rebuild its prosperity." He ended by apostrophizing, in a loud voice of conviction, *"Peuple français, grand peuple! Fierté, courage, espérance! Vive la République! Vive la France!"*

1959

January 14

No French chief of state has ever been voted for so much or so often as Charles de Gaulle—first in the September referendum, second in the general presidential election, and then, for the third time, in the presidential election by the special electors—the newly invented founding formula of his Fifth Republic being triply tested to make sure it will stand. On Thursday's Eurovision, or European television hookup, covering the procedure in the sumptuous chandeliered ballroom of the Palais de l'Elysée by which the supreme office of the state was transferred from the shoulders of the outgoing President Coty to those of the incoming President de Gaulle (seen for the first, and probably only, time dressed in a civilian cutaway coat), it was to be noted that once more the vote played a role—its last one. It came when an unfamiliar little white-bearded old Frenchman, who looked like a figure left over from the early Third Republic of a lifetime ago but who turned out to be M. René Cassin, head of the provisional Constitutional Committee that created last September's new constitution, advanced to the microphone. In a voice trembling with a sense of history, he read aloud the number of ballots cast for the victor and the two losers in the special electors' voting (when, until almost the last minute, it had been supposed that de Gaulle would be the unique candidate). The votes for the General, M. Cassin said, were 62,394; those for the Communist candidate, M. Georges Marane, were 10,355; and those for the Radical Socialist candidate, M. Albert Chatelet, were 6,721. In consequence, he then declared, the provisional Constitutional Committee—and here his old voice rose—"proclaims General de

Gaulle President of the Republic." With that phrase and at that precise moment—twelve minutes past noon of a sunny, cold Paris January day—in the eyes of the law and of French history Charles de Gaulle became the first Chief Magistrate of the Fifth Republic, which he created. Also at that moment began the twenty-one-gun salute fired by cannons placed beside the Seine to honor the arrival of a new head of state.

After General Georges Catroux, Grand Chancellor of the Legion of Honor, standing on tiptoe, hung around the giant de Gaulle's throat the great golden collar of the Grand Master of the Legion of Honor, founded by Napoleon—the badge of office with which all the French presidents are invested, and which they wear only at this ceremony—Coty recited his brief speech of welcome and adieu to the new incumbent of the palace. For this, he gracefully used Pascal's famous conjunction of "the grandeur of the establishment" and "the grandeur of the person." De Gaulle's speech of thanks was grave, impassioned. At first, he looked close to manly tears. He said that it was his duty to represent the interests of the nation and the *Communauté,* and even to impose them, since "these are my obligations." He added, in clear warning, "In this I shall not weaken—I bear witness to that in advance." After calling for the same aid that was given him last spring during national danger and is needed again "as the horizon brightens with our great hopes," he ended by crying, *"Vive la Communauté! Vive la République! Vive la France!"*—the new *vivat* of the Fifth Republic. De Gaulle's final shout of *"Vive la France!"* seemed to come from the very depths of his lungs, closest to his heart.

And now what? De Gaulle's government supporters, exclusively Right and Center, add up to a parliamentary majority of 88.8 per cent. Parliament will have its first brief sitting tomorrow, to hear, debate, and certainly accept the government's program—the first debate it has had since early last June—after which it will go home again until the last week of April. It is not clear what it will do when it does return, except mostly continue to say yes, which it was largely elected to say by the French people. (Were it to prove fractious and start saying no or perhaps, de Gaulle, by his constitution, has the power simply to dissolve it after one year.)

As the French intellectual Left analyzes it, the entire Left lost its strength and meaning by its démodé class-war ideology over the past two or three years, falling out of step with the Fourth Republic's surface prosperity, rising living standards, rising wages, and social-service handouts paid for almost entirely by the owner class, behind all of which were menacing state financial problems that were too much for mere dialectic to solve in a democracy. As for the Communists, the intellectuals think that Khrushchev, as much as anyone, cut them down from the biggest to the littlest French party, first costing them their belief in Moscow's infallibility by the revelations on Stalin, then undermining them with the bloody purges, followed by Budapest and, finally, even by Boris Pasternak, whose "Le Docteur Jivago" (as it is called here) was the best-seller of 1958—and not only among the bourgeoisie. Since the Socialist and the Communist Parties, which were sworn enemies in Parliament, have lost their influential major-party seats, it is now feared that they will make their real opposition to de Gaulle outside Parliament, and together—in strikes.

January 28

"At last we have a retrospective exhibition, at the Galerie Charpentier, of a hundred and thirteen pictures worthy of the highly original scenic art that was Maurice Utrillo's—at least up to about 1925," wrote one candid critic of this newly opened and biggest art attraction of the midwinter Paris season. For once in a modern French art show, the pictures have almost all been lent by French collectors, Utrillo being the sole famed School of Paris artist whom the French immediately appreciated and purchased, leaving the pictures of the half-dozen others who in various ways shook the art world to be admired, bought, and taken home by foreigners. As if to give a sharp educational correction to the public's sentimental affection for any Utrillo picture at all that is of Montmartre, is smooth, and is white, the opening fifty Utrillo canvases featured by the Charpentier are his early blue-and-green gems, as roughly faceted as if by Sisley himself, whom Utrillo then greatly admired, and painted in country towns or back gardens, like the Montmagny series. Furthermore, two-thirds of the pictures in the show fall strictly within the brief twenty-year period, from 1905 to 1925, when, as the catalogue preface makes clear, the tender genius of his painting was at full strength, with the apogee, the infinitely appreciated White

Period, coming between 1908 and 1914. Then, after 1925, began the decline, which led him first into honorably mediocre and finally into merely naïve painting. In 1923, his devoted mother, Suzanne Valadon, the painter, former Renoir model, and ex-circus performer, who had vainly been trying to keep her son off drink since he was a boy of ten (his father, if he was the man she supposed he was, had himself been a hereditary alcoholic), actually bought a lonely feudal château in the Ain for Maurice—with his money, of course, for by now he was a big success, who had exhibited at Paul Guillaume's. Here she lived with him for two years as one of his guardians. He was given a bottle or so of red wine, like good-conduct pay, only after he had finished his day's painting stint for his art merchant. Even in that rural spot, his work consisted of transforming and illumining scenes shown on the local postcards—a technique that had become basic to him. Maybe in that half-furnished nightmare medieval castle he became permanently disoriented by nostalgia for his cardboard vistas of Montmartre streets, leading around corners to nowhere, or perhaps luxury had become his most destructive imprisonment, whereas misery and poverty had, earlier, left his vision free. In any case, his rare emotional essence as an artist, always haunted by walls (which he put on canvas perhaps in memory of his many captivities), began deserting him, leaving only his manual talent, which his years of inebriation had never blurred. Between his teens and the age of forty-one, he had been, as a dipsomaniac, confined ten times—once for three years—in private clinics or Paris lunatic asylums, and always he had painted as well inside them as outside them. Until about 1946, when he was sixty-three, he continued painting with some coherence—mostly inept flower pictures, in which dahlias look like carnations—working under luxurious surveillance in a Paris mansion with barred windows. He had become a rich man who could buy anything for himself except liberty, for, as a famous national figure, he could not be allowed freedom. The government made him an Officer of the Legion of Honor, he was allotted an entire room for his paintings at a Venice Biennale exhibition, his white pictures of Sacré-Cœur—and even the counterfeit ones—were loved over the Western world as the essence of Paris. But he himself was like a guttering white candle, kept indoors. He died only in 1955, aged seventy-one, in a rich villa in suburban Vésinet (equipped with a private chapel for his inchoate prayers to Jeanne d'Arc), where he was watched over by his bossy wife, a Belgian banker's widow to whom his dying mother had married him.

The catalogue gives Utrillo's formula for his famous white paint—white of zinc, ground plaster, and glue, applied with a palette knife. But it gives no explanation of the hallucinatory snowfalls with which he later covered Montmartre's streets—snow, after all, being a great Paris rarity, except, perhaps, to a painter with his own favorite recipe for white paint. Among the pathetic memorabilia in the exhibition are illustrated poems, scribbled on brown paper in Utrillo's schoolboy hand, denouncing the Montmartre children and housewives who tormented him in his public drunkenness until they drove him off the streets, where he painted from real life, and into his lonely room and his enforced practice of painting from postcards—*"Une population profane,"* he fulminated, *"une voyoucratie,"* or government by hooligans. This enormous Charpentier retrospective implies that for the next years in Paris there will be no sizable Utrillo exhibitions. The art critics, as if in farewell, gave Utrillo a garland of their talented praise, such as "The painter of poverty and abandon, he transfigured the most wretched scenes by the tenderness that lighted and colored his heart." And "Tender and conscientious, with delectable diversity, he evoked the pathos of Paris and the anonymous soul of the suburbs."

The production, at Charles Dullin's picturesque old Théâtre de l'Atelier, on its tree-decked square behind Pigalle, of "La Punaise" ("The Bedbug"), by the noted Soviet poet Vladimir Mayakovsky, has given intellectual disappointment but childish ocular pleasure since its opening last week. Anticipatory interest had been aroused here by the fact that this would be the first time such a specimen of Soviet humor had been played in Paris, and also by the fact that the play is being currently presented at the Satire Theatre, in Moscow. Probably too much was expected of it because so much was already known about Mayakovsky. Paris has an active Communist press-and-book center, which, willy-nilly, keeps us abreast of some Moscow cultural news. Mayakovsky's poems, in beautiful translations, have become well known here, as has, recently, his youthful friendship with Boris Pasternak. Also familiar was Mayakovsky's early, stimulating Moscow career as a so-called Futurist, or avant-garde leader in art, poetry, and ideas; then as an ad writer and a designer of striking posters for various Soviet state enterprises; and, just prior to 1930, as a leading journalist on the *Komsomolskaya Pravda.* Indeed, almost everything was known here about Mayakovsky except why he put a bullet through his heart in April, 1930. However, this followed the 1929 production of "The Bedbug," in Stalin's time, at the Meyerhold Theatre, where it was a failure, plus violent attacks, in March, 1930, on the Meyerhold production of his play "The Baths," which was such an assault on Soviet bureaucracy that it had to be withdrawn, as Mayakovsky himself withdrew from life the next month. Both plays were successfully revived in Moscow a few years ago, and, in the Khrushchev thaw, they have been in the Satire Theatre repertory ever since. As far as is known here, the Atelier's is the first production of "The Bedbug" in Western Europe.

The idea—at least as many of us had imagined it in advance—is extremely funny, capable of being cruelly comic and loaded with the deadly satire that can come from anachronism. It concerns a post-revolutionary lout named Prisypkin, who deserts his laboring-class comrades and milieu to marry a fat manicurist. In the wedding party's drunken revelry, the house burns down. All but one of the revellers are pulled out, and after the firemen have flooded the site with water, it turns into a block of ice. Fifty years later, Soviet scientists vote to thaw out their latest great scientific discovery—a subhuman in ice—who turns out to be Prisypkin himself, not dead but frozen, and who is soon restored to life with everything intact, including a bedbug that had been on the wedding party with him. By now, both he and it are sociobiological forms unknown in advanced Soviet culture, he being classified as a *Petit-bourgeoisus vulgaris* and the insect as a *Punaispa normalis*. The bug is seized for observation by the zoo. The thawed lout, reeking of vodka fumes in a Soviet Union so pure that his breath makes the Comrades sick, infects all society, which breaks out with a passion for drink, for dancing the Charleston, and for acting so contrary to Marxism in general that even the dogs start walking the streets on their hind legs (which a couple of them obligingly do on the stage). The finale shows Prisypkin in a cage at the zoo, where he has been put to furnish food for the bug; suddenly he sees the theatre audience, whereupon he shouts happily at it, recognizing the audience to be like himself—made up of unfrozen human beings—and the play is over.

Even in such bald outline, one can see how good the play sounds and ought to be. Well, it is not much good as played here, and no good at all as a slashing satire. There are thirty leading characters in it, and about fifty extras, who play double walk-on roles in what looks like a jolly Chauve-Souris-style production, with treadmills onstage to imitate the passing of crowds, scenery that moves, and so forth. The wedding party's drunken revelry is the drunkest-looking and longest-lasting drunken scene in the memory of Paris theatre-goers—a banging, falling-down affair of nothing but naturalness and pantomime, without a word spoken, which goes on for perhaps twenty solid minutes. Many of the spectators laugh uproariously; many yawn in silence. The *Figaro* critic said of "La Punaise" that it had bored him to the point of anesthesia. One has the feeling that Mayakovsky's script has disappeared from the Atelier production under a muscular mass of interpretive pantomime, though the Atelier producer, André Barsacq, is himself a Russian, it turns out, who translated the Mayakovsky opus and thus should have known what he was doing. It is strange that this so-called satire on a new modern society by a brilliant, idealistic mind so acute in its powers that it could not bear living in that society any more should seem a less penetrating comment on the Russians' perpetual search for personal liberty than that given us years ago by the fairy-tale ballet "Petrouchka," where the struggling characters had only the tragedy of being dancing puppets, and their dictator merely the tyranny of a magician.

March 11

The retrospective exhibition called "The Twenties: American Writers in Paris and Their Friends," which has just opened in the St.-Germain-des-Prés quarter where they lived, is an act of recrudescence. It restores to life that memorable, far-back decade when talent and faces were fresh, when the young expatriates came here in numbers and in a united coincidence and founded what became the new contemporary school of American literature. The exhibition is under the aegis of the cultural wing of the American Embassy and is being held in the new United States Information Service Cultural Center on the Rue du Dragon, the narrow, short street just beyond the Brasserie Lipp and across and down from the Café des Deux Magots, those focal points in the old days. The exhibition's six hundred items, many rare, owe their presence almost entirely to the fact that Miss Sylvia Beach, of the Shakespeare and Company bookshop, on the Rue de l'Odéon, has never lost anything or thrown anything away. So now here it all is, gathered together with a few memorabilia from others of the epoch—the scribbled, jocose notes, the corrected page proofs, the photographs, the snapshots, the first editions dedicated "to Sylvia, with love" from Hem, Dos, Djuna, E.E., Ezra, Thornton, Gertrude, Scott, and others. As source material of literary and documentary importance, it is a miraculously complete show, with the first editions of the expatriates (often printed here in France on private presses by other expatriates, because no New York commercial editors had the faith to print them), and with examples of all the vital, struggling literary magazines from everywhere—splendid silken ragbags, sometimes backed by the rich—for which, in the early days, the poor-of-purse writers gladly wrote without remuneration, just to see their wonderful words in print. It is also a clearly, handsomely, and imaginatively arranged exhibition, using glass cases set upright against the walls, so that what is in them is easily visible and legible, with entire cases devoted to an outstanding writer—including his letters, manuscripts, photographs, snapshots of family increases, bullfights, fishing trips, and so on—so he can be studied all at once.

The expatriates' literary clubroom, the Shakespeare shop not only sold books in the English language but also rented them out as a lending library. Miss Gertrude Stein was Sylvia Beach's first subscriber, and early wrote a poem called "Rich and Poor in English to Subscribers in French and Other Latin Tongues," which Miss Alice B. Toklas typed off—there is a typescript copy in the show—and mailed to friends to encourage them to subscribe also. Robert McAlmon, who ran Contact Editions, a private press, published Miss Stein's enormous "Making of Americans" (shown along with one of her corrected page proofs), because Hemingway wanted him to and because no publisher in America would print it, and also printed Hemingway's first book, a small blue paperback called "Three Stories & Ten Poems." William Bird, who had the Three Mountains Press, published Hemingway's "In Our Time," a tall yellow book whose cover was ornamented with disjointed cuttings from newspapers, printed in red—phrases including "two billion dollars," "guidance from God," and "learn French." In the Hemingway showcase is a jovial note to "Madame Shakespeare," meaning Miss Beach, in which he remarked, "To hell with the book." In Scott Fitzgerald's showcase is a copy of "The Great Gatsby" dedicated to "Dear Sylvia, from Harold Bell Wright," which he crossed out, signing his own name. Bird also published "XVI Cantos," by Ezra Pound, who had brought Joyce and his "Ulysses" from Trieste to Paris in that almost willful concentration here of talented outlanders who were having trouble being published. Because English typesetters refused to set type for the bawdy parts of "Ulysses," Miss Beach and her Shakespeare shop became its publisher, the printing being done in Dijon by Frenchmen, ignorant of what the English words meant. In the Joyce showcase are order forms for the 1922 first edition of "Ulysses" sent in by Lawrence of Arabia, who ordered two copies of the most expensive edition—on Dutch paper, at three hundred and fifty francs—and by Yeats and Gide, who each ordered a copy of the cheapest edition, at a hundred and fifty francs. Also shown are some of Joyce's corrected page proofs. As the publisher of "Ulysses," Miss Beach became both martyr and heroine when its detractors and admirers began congregating, over the years, in her shop. Near the Joyce material, flanked by dashing photos of Margaret Anderson and Jane Heap, is the famous 1920 autumn number of the revolutionary *Little Review* in which they announced "our arrest" for having published sections of "Ulysses" in New York.

In the exhibition's photograph section, entitled "Portraits of a Generation," and mostly taken by Man Ray and Berenice Abbott, everyone is present who was attached, in one way or another, to the Shakespeare shop: T. S. Eliot; Eugene and Maria Jolas, of the intransigent magazine called *transition;* Archibald MacLeish; Katherine Anne Porter; Allen Tate; Virgil Thompson; Djuna Barnes, in profile in a cape; Kay Boyle, front-face in a beret; Dos Passos; Bryher; Arthur Moss, of *Gargoyle;* Louis Bromfield; Cummings; Sherwood Anderson; Wilder; Nathanael West; Caresse and Harry Crosby, of the Black Sun Press; Nancy Cunard, of the Hours Press; Alexander Calder; George Antheil; Edmund Wilson; Mary Reynolds; and others—plus the French writers and artists drawn to this new fire of foreign talent blazing in their own city, among them Gide, Valéry, Larbaud, Schlumberger, Chamson, and Marcel Duchamp.

In one corner of the exhibition room, the walls are covered with a photo-montage of the façade of the old Dingo café, in Montparnasse, where the St.-Germain talent spent many of its nights over the years. Real café chairs and tables are placed in front of the montage, and there, late in the afternoon of the exhibition's opening, sat Miss Alice B. Toklas and Thornton Wilder in literary reminiscence, while behind them a pianola beat out the rhythms of Antheil's "Ballet Mécanique," the shock music of that decade.

March 25

The highest skyscraper in all Europe—fifty-five stories—whose work-in-progress name, if you can believe it, is Antigone (daughter of Oedipus, who became his guide after he put out his own eyes), will figure as the towering center of the controversial complex of ultramodern buildings that are to replace the shabby old Gare Montparnasse and incorporate the new one. The project, which comprises four separate sections, is to start this June, it has just been announced, and will be finished, *Deo volente,* in 1964. This will constitute the greatest Paris urban project since that of Baron Haussmann, back in the eighteen-fifties and sixties. The skyscraper is to be the railroad station's hotel, with a night club on top. It will rise to more than half the height of that Paris landmark and trademark the Tour Eiffel—also, in its work-in-progress state, resented by the populace, who then, as now, wanted nothing ultramodern. It will be visible from the Place de la Concorde, the Place St.-Germain-des-Prés, the Esplanade of the Invalides, and the terrace of the Palais de Chaillot, and thus will change the familiar, loved silhouette of Paris. The hotel, approached by a *parvis* facing the Place de Rennes, will contain a shopping center, a cinema, a press club, a swimming pool (maybe), and a hall for public meetings. Beside it will lie the railway-station section, its entrance flush with the sidewalk, but with the tracks unfortunately nested nearly twenty-five feet below, which means stairs and suitcases, mixed, for the travellers. The outgoing tracks may have to be covered with a garden as far as the Boulevard Pasteur, for the sake of the nineteen-story buildings on either side of them—one for offices, one for apartments, with artists' studios on top. The triangle remaining on the Boulevard de Vaugirard is to be a green park, with a subterranean parking lot for fifteen hundred cars. Part of the project's cost will be borne by the Ville de Paris and by the S.N.C.F., the nationalized association of French railways, but most of it will be met by private investment. To those Parisians who outspokenly regard the fifty-five-story skyscraper as aesthetically unappealing, non-native, unnecessary, and bound to throw the architectural equilibrium of the famous Paris skyline out of plumb, the Council of Buildings of France gave a tacit rebuff last week by voting final approval of the entire project, thus officially accepting a new architectural future. The Council's opinion was: "Paris cannot afford to lose herself in her past. In the years to come, Paris must undergo imposing metamorphoses." *C'est bien triste.*

April 7

For the English-speaking colony here—provided it could follow the Irish accents—the treat of the current Théâtre des Nations festival has been the London Workshop's uproarious performance of Brendan Behan's hit "The Hostage." On Monday, its last night, Behan, who had been on public view drinking Pernods in the theatre bar, unexpectedly joined the company as it was taking its final bow onstage, and burst into a long, ponderous Irish jig, stamping his huge feet in inebriated rhythm, while the audience clapped to help him keep time. As symbol, explanation, and integral proof of his play—a talented, showoff Irishman, gay and belligerent in his drink, slovenly and confidential in his shirtsleeves and suspenders, the Gael himself, unique, theatrical, spirited, his national essence a form of patriotism unconnected with the rest of the world—he was as good as his play, and just like it. "The Hostage" takes place in a shabby semi-brothel and I.R.A. headquarters in Dublin, and is animated by motley characters—including one lunatic, dressed in kilts like Brian Boru—who burst into bawdy or anti-British songs, and drink and quarrel constantly, all with a rich gift of the gab that is their only wealth. The hostage himself is an English soldier boy, held against the life of an Irish soldier who is due to be hanged at dawn in Belfast Castle. The Tommy is shot anyhow, by accident, in a midnight brothel raid; comes to life again; and joins the others in singing "The Bells of Hell Go Ting-a-Ling-a-Ling," the finale. It is all mad—a superb, violent, civilized evening's entertainment.

Spring has finally come to Paris, but this year in a different style, almost with a new method. In the Tuileries gardens it came cautiously, tree by tree. On Friday, there was a tree or so that sprouted green leaves; another scattered dozen burgeoned in Saturday's heat; and on Sunday the pattern was set by verdure everywhere, with the giant blue Paris pigeons bolting above it, two by two.

April 22

The floral treat of the spring, of the year, and maybe of all time in Paris will be offered this Friday (for ten days only), with the opening of the first, and perhaps the last, of the Floralies Internationales ever to be held in the French capital. The greatest flower-and-shrub show in the city's history, it will be displayed in the city's newest, most imaginative twentieth-century construction, Le Palais de la Défense, only recently completed as one of the loveliest exhibition halls of our time. It rises just outside the city limits, at Puteaux, on the Rond-Point de la Défense, where Paris vainly made its last stand in the war of 1870 against the invading Germans, and is the pilot building of the new Paris that, in accordance with the city's modern urbanization plan, is starting to stretch west with skyscrapers and vertical communities that will soon lead to and connect with St.-Germain-en-Laye, the royal town where Louis XIV was born. In appearance, the Palais de la Défense is a pure and also a practical fantasy. It consists of a ribbed roof of pre-stressed concrete that looks like a billowing white parachute slit below into three parts, with each part coming to rest on its pointed tip upon the earth, making a three-faced building, walled in on its arched sides by glass only—the largest and the airiest-looking exhibition hall in all Europe, now filled with Europe's greatest potted garden of flowers and bushes in bloom. Even as seen in preview, the combination of the floral gifts of nature and the architectural gift of contemporary man makes this an exhibition to endure for long seasons in one's eyes and memory.

It is becoming increasingly clear that France's Fifth Republic is like no other she has ever had—in many ways a blessing—and that in one respect, despite the now unquestioned integrity of General de Gaulle's republican sentiments, it strangely resembles a monarchy. Among the French, at least, when a man, whether as President of the Fifth Republic or as a sovereign, holds power of a certain exceptional sort, from which nothing except death or revolution can remove him within a given period of time, his national popularity (or the contrary) becomes the pulse of his governing. About de Gaulle's popularity there is constant public and political curiosity (Is it going up again? Has it gone down? Feverish? Cool?), as though it were a kind of emotional temperature registered on some mostly invisible thermometer of public opinion that the public itself and the newspaper editors are constantly trying to calculate. This was not true, of course, of France's normal political leaders of the Third Republic—the historically notable Premiers like Clemenceau, for example, or Briand or Blum, or even Daladier—for whom the important thing was their popularity with the other politicians in Parliament, and who were not constantly engaged in a nationwide popularity contest, as de Gaulle is now.

Because de Gaulle is today's sovereign personality, his stately visits to the provinces, such as the one he has just made, in the pouring rain, to Burgundy, to the industrial center of France, and to the agricultural southwest, are of great informative importance. On the whole, he was unpopular everywhere with the war veterans, who failed to turn out with their battle flags to receive him, because he had cut their pensions. He was unpopular with the little winegrowers around Mâcon and through all that region, because his government has doubled the tax on inferior wines, continued its anti-alcoholism campaign, and followed the previous government's policy of ordering vines of inferior grapes to be torn up in order to make way for edible crops, in a modern agricultural plan. He was, however, extremely popular with the winegrowing farmers in the Burgundy regions of *les grands crus,* which still sell splendidly from the hills of Chambertin, Vosne-Romanée, Clos de Vougeot, and Beaune. He was unexpectedly popular with the majority of the thirteen thousand workmen at the great Creusot steel works (formerly the old Schneider works, which used to manufacture cannon and are now making machinery for Russia and Communist China), where the men who cheered him drowned out the few who whistled against him. At the Dunlop tire factory in Montluçon, which is a fief of the C.G.T. labor union, dominated by the Communists, the Party had ordered the men to lay down their tools and fold their arms as a sign of displeasure during the General's visit. The men stopped work, all right, but for the most part to shake hands with him. In the ancient village of Cuisery, near the cathedral town of Tournus, he made an extra-popular brief visit to the Cuisery chapel to see the tomb of a 1704 ancestor of his—the de Gaulles having originally been a Burgundian family, starting with a certain Captain Gaspard de Gaulle, given special privileges by the King in 1581. This was an extra-popular part of his tour simply because the French respect lineage. In one town, he sneezed into the microphone, trumpeted into his handkerchief, and confided hoarsely to thousands, "I've had the grippe. But it's much better." Twice, he was too hoarse to add his loud, off-pitch voice to the singing of the "Marseillaise." He repeatedly spoke outdoors in the rain, was hatless, coatless, drenched, good-natured, optimistic, and humanitarian. The theme of his speeches everywhere was the same, and could be summed up as "Not for the good of France alone but for the good of all men, let France go forward!" To the Montluçon workmen he said, "The world is round. There is only one world. All the men of the world are men like you and me. [Applause.] I have confidence that this year will see the beginning of that fraternal [world] organization for which France will set the example." Deliberately, he chose to stop off at Vichy, former government seat of Maréchal Pétain, whose name he did not mention. Instead, with great dignity, he said in a downpour to the thousands of umbrellas of the Vichyssois surrounding him, "It is, I will say to you in confidence, rather moving for me to find myself here officially. History makes its links. Whatever the changes, we remain one people, we are the real, the unique French people." Throughout the trip, his private car flew his private flag—the French tricolor pennant struck with the cross of Lorraine—and throughout the trip his popularity as the great, patriotic, highly educated, eccentric, and powerful President of the Fifth Republic was on the whole undeniable, and at times even touching.

May 20

The great surprise that has contributed to the de Gaulle Republic's quasi success, so far heavily qualified by many losses of invisible values, is the altered psychology of Charles de Gaulle himself. It seems thrice altered, actually—from his original rigid, patriotic tactlessness with President Roosevelt and Prime Minister Churchill to his sudden flight from responsibility when he was head of the provisional postwar government, and from there to his present metamorphosis, after his isolated years at Colombey, reading and writing history but not living it, except by his sibylline prophecy that France in real disaster could call only on him for salvation, a combination of unlikely and alarming eventualities no one had dreamed of till they came true. This time, he came to power a changed man, an obstinately responsible, superorganized elderly chief and educated patrician, so superior a French figure that even his dreams for a revived French *mystique* of mission and glory seemed merely gentlemanly visions. With all these, he brought new wisdom and grace—his Machiavellian ambiguity giving him the magic ability to function with Frenchmen of opposed views, which he amalgamates.

It may be recalled that millions voted for the General out of sheer fear of nothingness. Thus, at least once—and it may be that this was his only miracle—he united the untouchables of French politics: those extremists in a French republic's left and right who over the decades have regarded each other as doctrinally impure, and in their quarrels have cut France to pieces. There is already far more worry about who will take over the power inherent in the Fifth Republic's strong Constitution once de Gaulle is retired—or dies, he being no longer young—than there was about the dying Fourth Republic and its Constitution's anemia. How this Fifth Republic can democratically function with a bobtailed parliament that rarely meets—it is almost unnoticeably meeting now, with little power to debate and almost less to legislate—nobody yet knows. France, as plenty of its citizens now realize, is living in a new kind of republican world, with new and fundamental dangers to match.

June 3

For a month, Paris movie fans and critics have been debating the problem posed by Ingmar Bergman, the Swedish film director who is now ranked by those who reverently admire his works—and, perforce, even by those who don't—as the dominant world cinematographer today. Technically, he is an imagist who practices *Expressionismus,* and he has made twenty movies since 1945. Most leading film critics (far more important to Paris entertainment than mere theatre critics) have lauded him with eulogies of a fervor never heard of here before—at least not in relation to a row of painful movies, in an incomprehensible foreign tongue and with inadequate French captions, made by a man who has "a Swedish preoccupation with loneliness, death, and eroticism." The films shown here, at select neighborhood houses, have included "Smiles of a Summer Night," the glacially gay one about a middle-aged man who has a new young wife the same age as his son by a first wife; "The Prison," about prostitutes in one, and infanticide; "The Seventh Seal," about a medieval knight amorous of death; "Summer Games," about swimming in the fiords, and a young lover who dives headfirst onto a rock and dies; "Threshold of Life," which is set in a maternity hospital; and his latest, "Wild Strawberries," which is being called his masterpiece. The leading role is wonderfully acted by the elderly Victor Sjöström, an old star from the Malmö Stadsteater (which Bergman has also directed). He plays a venerable doctor who, at a ceremony honoring him as his city's finest citizen, realizes, in dream sequences and memory flashbacks, that he has killed love throughout his life. The "Strawberries" review by Claude Mauriac, the sincere, perspicacious movie critic of the weekly *Figaro Littéraire,* was an illuminating example of the Paris fervor. Among his phrases were "Bergman, a great, a very great artist . . . cinematographic miracle . . . the cruelty equals the poetry . . . humanism as well as harmony . . . cinema on a level with the great modes of noble creation." He touchingly added, in a personal address to his readers, "If you see only one film in six months, this is the one to see."

Bergman's direction of the Malmö Stadsteater production of the homonymous Hjalmar Bergman's play "A Saga" was considered harsh and cold when it was recently given here, in Swedish, at the International Theatre Festival. Since the theatre is Ingmar Bergman's real vocation, with moviemaking only his hobby, each May, when the theatre season ends, he enters a hospital in a state of collapse and stays six weeks, writing his next film, which, during Sweden's long summer days, he shoots in a furious two months of concentration. He hates critics; when he was here for "A Saga" and the Paris movie critics admiringly asked him why he made his films the way he did, he coolly said, "To mystify." According to a non-French report, he "is a character straight out of Strindberg, neurotic, insomniac, hypochondriac . . . neither smokes nor drinks; rarely shaves; goes about in corduroys and an old brown sweater . . . and once threw a chair through the control-booth window at a sound mixer who had bungled a recording." He is the son of the old King of Sweden's chaplain.

July 2

In this June season, the most talk and critical interest have, for once, centered on entertainment that everybody can afford—the three French films that, one way or another, were top winners in this spring's international Cannes Film Festival, all three having peculiar qualities, and peculiar men and histories behind them. The Golden Palm winner is a French film that was made in Rio de Janeiro, and in Portuguese (there is also a dubbed French-language version), during last year's carnival down there. It is called "Orfeu Negro," and it is an emotional shantytown, samba version of the antique legend of Orpheus and Eurydice, acted by Negro natives of Rio. Its hitherto unknown director—in all three cases, it is new directors, and not the actors, who have become the stars—is Marcel Camus (no relation to the novelist), aged forty-six, who had directed one film before in his life, and that one not noteworthy. He was so short of funds that he could buy only a one-way ticket to Rio to make this film, which he had determined upon, and was so heavily in debt by the time he started his return trip to France that he almost missed being allowed to take his unpaid-for sound track back with him. The seduction of his film is its naïve quality, emanting from its untrained Negro actors, who are like local shadows repeating the motions of a great civilized myth. The main site of its drama—the high mountain plateau above the sea and the city—affords magnificent scenic views from the ramshackle cabins where the protagonists dwell. The Eurydice is a pretty, light-colored girl named Marpessa Dawn, reportedly a Puerto Rican who has lived in both Philadelphia and Rio—but this is not a film in which civil identities count. During the day, Orpheus, a big, handsome black man, is a streetcar motorman down in Rio, but at night he takes on his humble musical role by strumming his guitar in his plateau hut; Eurydice, who becomes his beloved, is a visitor in her cousin's cabin, next door. The big scene of the film, for which it was obviously made, is the first carnival night, with what looks like miles of Rio streets covered by garishly costumed natives doing the samba with hypnotic fervor to the insistent pulse of multiple brass bands. Among the crowds is a mysterious male figure, disguised in a carnival costume of death, who pursues Eurydice into an electric powerhouse, where Orpheus, seeing her frightened face, and intent on saving her, accidentally touches a switch and she is electrocuted—certainly the most innocent, mechanized metamorphosis of the old legend yet contrived. Technically, the film is often rickety, but of the three great Cannes triumphs it is the one the French love most and argue about least.

The dominant Cannes film, which actually won nothing there except a general agreement that it is a passionate, troubling chef-d'œuvre, is "Hiroshima, Mon Amour," which was made in Japan and is the second long film directed by the almost unknown thirty-seven-year-old Alain Resnais. It opens on the embracing nude torsos of a Japanese man and a Frenchwoman in Hiroshima, after the couple's first, and only, night of love. He is, it turns out, an educated bourgeois; she is a French actress who has just made what one gathers is a pacifist film about Hiroshima. However, details are left submerged in the actual movie; the complex pattern of its emotions and ideas forms its substance. To judge by what has been written about it here—and it is the most written-about, talked-about European film since the war—every movie critic has had difficulty giving an adequate description of what happens in it, because attempting a verbal translation of its multiple photographic facets would be like trying to summarize the lyric speed and complex structure of some new piece of sensuous, emotional poetry. The film's main theme is what the French, in a passing fit of revivalist enthusiasm for a perfect tragic love, like that of Tristan and Isolde, now call *l'amour passion*—an ideal, doomed love that, because of its magic brevity and the forces working against it, marks the lovers' lives and is romantically preserved in memory across the earthly distance that ever after separates them. It is this kind of love that unites the Japanese man and the European woman, condemned to be separated that very day by her return to her own land. Like the film within it, "Hiroshima, Mon Amour" is a film of pacifism—a pacifism that stems, basically, from the two politically experienced brains that produced it, turning it into artistry. Whatever director Resnais's precise political attachments may be, he is clearly angry in a world of constantly increasing atomic bombs. The well-known novelist Mlle. Marguerite Duras, who wrote the film's scenario and dialogue, was, she declared in a recent interview, a Communist Party member until she was thrown out for the heresy of her views on Budapest, and she is certainly anti-bomb where the American devastation of Hiroshima is concerned.

The theme of the last of the three Cannes films is international only in that it concerns a delinquent boy. A hard, clear, excellently created movie about a thirteen-year-old Place Pigalle lad, it is called "Les Quatre Cents Coups"—a French idiom for what we would call "a bad lot"—and was directed by twenty-seven-year-old François Truffaut, who is anything but unknown in Paris moviemaking and movie-fan circles. As a precocious, authoritative critic, writing of late mostly in the weekly *Arts,* he has for several years been brilliantly insulting about some of the most successful and popular French moviemakers. A member of what the French call *"la nouvelle vague,"* or "the new wave," which means any young French people who are sick of old methods and ideas and are trying to utilize their own new conceptions, he has been the leader in disseminating the revolutionary notion that French films can be intelligent, even truthful to life, and still make money, citing as proof such directors, of varied nationalities, as Renoir, Hitchcock, Rossellini, Bresson, and Ophuls. He so impertinently, and justifiably, attacked the boredom that recent obsequious Cannes Festival decisions on prizes have begotten that for the past two years the Festival directors shut him out as a critic. This year, not only did the French selection committee choose his film to represent France but he saw it win the second important prize at Cannes—that for the best direction. In speaking of this film on juvenile delinquency, Truffaut, who is an explosively candid fellow, told the press that he himself had been a delinquent and had been put in a reform school, so he knew that his film represented real life, and he added that he thought it intelligent, and had every reason to expect that it would make money, which it undoubtedly is already doing in France. His boy actor is, in private life, merely a schoolboy named Jean-Pierre Léaud. In the film, he is plunged into a Parisian situation that is both sinister and commonplace. His parents quarrel in bed every night, and one day he sees his mother kissing a strange man in the shelter of the steps of a Métro station; he then discovers that his mother's husband is not his father, that she has been tart, and that he himself is illegitimate. The rest is classic: the boy runs away; steals, because he now has an obsession to go see the sea and needs money for the trip; is arrested, at his bewildered parents' request; is taken to jail, along with prostitutes, in the police wagon (the glitter of his tears and the corresponding twinkle of Pigalle's hedonistic night lights in the rain are a brilliant invention); and is sent to a normally brutal country reform school somewhere in western France. (The school psychiatrist's only questioning of the boy—exclusively on sex—is a stunningly good scene.) The boy escapes. Then follows a remarkable sequence that shows him running through the countryside, from one kind of France into another, from fields into hills into marshy plains, always breathless, as if escape took forever. The picture finishes when he sights the sea, runs into the water up to his ankles, turns, gazes at the landscape as if he were now insulated against pursuit, and, lifting his expressionless black young eyes, gives a long, a very long, motionless look beyond, which falls on those who are looking at him—on the spectators in the theatre, on them as humanity, on the society they represent.

July 15

For superior vacation reading, two novels of exceptional quality have attracted unusual attention at the tag end of the publishing season, one of them, "Le Planétarium," by Nathalie Sarraute, having been cited as a chef d'œuvre of the subconscious—a rarity indeed in contemporary French writing. Mme. Sarraute was born in Russia, where her mother was a writer, her father a chemist, and his brother a terrorist who, after attempting to assassinate a grand duke, fled to Sweden disguised as a woman. Long a resident of Paris, married to a French lawyer, and mother of three grown daughters (one also a writer), she is today still foreign-looking in her way, with her black hair combed back, in the classic Russian style, above extraordinary eyes, as black as black cherries—an impressively quiet, observant, international-looking intellectual. She has always written slowly, she says, and has produced only five books in twenty years. The "anti-novel" has been her aim as part of the experimental literary movement here known as Le Roman Nouveau, or New Novel, of which you may at least have heard, since it has been talked about a great deal more than it has produced. Its only two notable public successes have been Michel Butor's strangely phrased, effective novel about a train journey, "La Modification" (called "A Change of Heart" in its American edition), which won the Renaudot literary prize last year, and now Mme. Sarraute's latest novel, already in its third edition. Perhaps it should be explained that the Roman Nouveau writers' general aim has been to break the classic fiction mold maintained by French genius over the past hundred and fifty years, in order to accommodate more properly the broken shapes and new psychologies of modern life. While these experiments have been eagerly followed by the Paris literary cliques, much of the intelligent Paris public has found them, on the whole, mystifying, and even downright bizarre—or, at any rate, hard to read. The young leader of the Roman Nouveau, Alain Robbe-Grillet, as his experiment, has developed an avant-garde neo-realism actually based on listing objects or things, which his British admirers have called his *chosisme*, or thingishness. His latest example of it a plantation novel called "La Jalousie"—is certainly astonishing. In the space of four pages, he lists the exact number of local banana trees—"the middle row, which should have had eighteen . . . had only sixteen," and so on. It seems worth citing the Robbe-Grillet formula for a New Novel if only because Mme. Sarraute's "Le Planétarium" is not in the least like it.

As Mme. Sarraute's title indicates, her novel involves the traceries of a set of people circling in their own orbits and gravitations of character like the spheres and planets, with their mutual attractions patterned against fatal collisions—an enlightening symbol of the complex mechanics of close human relationships today. The novel opens on an elderly aunt in her overlarge Paris flat; turns to the spoiled, ambitious, semi-intellectual nephew she has brought up, and to his wife, both of whom are living on the wife's affluent parents; and then to a passée famous woman novelist (a wonderful satiric, gaseous figure, as true and swollen with pride as a balloon), to whose literary salon the snobbish nephew craves entry, just as she, in her waning glory, craves to have recherché new young male faces ornamenting it. In progressive encircling relations between the old and young generations, the young couple obtain the old aunt's big flat through intrigue, and the nephew, moving into a higher stratum, is able to despise the pretentious old novelist as a distinguished bore—two accelerated new movements through social space that seem like a promise of his final airy freedom. "Le Planétarium" is really a novel about the avarice of the emotions at a particular level of bourgeois French hearts, to which Mme. Sarraute has added a modern sonority by using, in parts, the interior monologue, that terrible, echoing second voice of the present century. She seems the only one among the New Novel experimenters who appears finally to have struck her own style—intense, observational, and personal. Her book is to be translated into English by Mrs. Maria Jolas, one of the founders of *transition,* the famous prewar Franco-American experimental literary magazine here.

* * *

Another publication, which must be mentioned because of the grave shock it has produced, is a hundred-page booklet called "La Gangrène," composed of seven signed statements by Algerian students here—members of a government-banned rebel intellectual group—in which they describe their alleged torture by the Sûreté Nationale's notorious secret-police unit known as the D.S.T., against which the seven lodged useless formal complaints with French legal authorities last winter, after the events so cited. The booklet first came to public notice here just after its publication, in the last part of June, when a rousing article about it appeared in the independent afternoon paper *Le Monde;* within two hours police confiscated the remaining unsold copies of "La Gangrène" at the Editions de Minuit, the internationally known Resistance publishing office of the Second World War. A few days later, Premier Debré declared before the Senate that "La Gangrène" was "mendacious and dishonorable," and was written for pay by two Communists (an authorship that the Minuit office formally denied); and the Minister of the Interior actually filed with a Seine court a complaint against the publishers, charging "libel against the police." All these items, including accounts of the seizure and the contents of "La Gangrène," were published by all the Paris newspapers, but only *Le Monde* carried on a campaign for the booklet, declaring, "To call it infamous is not a substitute for the firm denial one would have liked to hear of the essential point: were the men tortured or not?" The tortures described by the seven Algerians (one of them the brother of the so-called Minister of Finance in the so-called Algerian rebel government-in-exile) allegedly took place partly in the cells of the D.S.T. in the Rue des Saussaies, about three hundred yards from the Palais de l'Elysée, the official residence of General de Gaulle. The booklet's seven signed statements are written in a matter-of-fact style, unhysterical though detailed, and are mostly similar. Stripped nude and flogged, the men were then laid over a bar, like fowls on a spit, and electrodes were applied to exposed parts of their bodies; two had to do knee bends while balancing the big Paris telephone book in their hands, and were beaten insensible when they failed. On page 97 of "La Gangrène," one Algerian describes the appearance of another after repeated beatings: "Only his protuberant eyes indicated that it was the face of a human being."

Perhaps the greatest shock produced by "La Gangrène" for millions of French comes from the realization that both the opposing prophecies about General de Gaulle before the referendum calling him to power fatally continue to be proved true—that he was France's only signal chance for recovery, and that though he is a man of the highest honor, always a patriot and now a liberal, he would to some extent be the captive of his Fascist followers. In the case of "La Gangrène," on which he has not said a word, he has also become the prisoner of his own silence. The French are now beginning to believe that while his painful, sacrificial silence at certain times is what is literally holding the straining Fifth Republic together, it is also dissipating part of its democratic citizens' hope and faith.

July 28

The only perceptible development in Montparnasse since the war is that, from the viewpoint of Paris history, hedonism, literature, and painting, it has slid downhill into quietude, losing its importance and crowds to the new night clubs and writers' cafés of St.-Germain-des-Prés. The latest proof is the razing of the notable old Café Rotonde, across from the Dôme, and the substitution on the same site of the newly opened La Rotonde Cinéma—a transformation performed with such an intelligent sense of nostalgia as to make the theatre the most extraordinary neighborhood movie house in Paris and a monument well worth visiting. Above the screen, an inscription by the poet Jacques Prévert greets the spectator—and if the spectator is left over from the old Montparnasse days and nights, it is enough to bring a tear of affection to the eye: "The names on these walls are those of persons of all sorts who haunted this place long ago. May they still be at home here as they used to be. This theatre belongs to their dreams and serves as their remembrance." Then, while the lights are still on and one sinks back into one's elegant rose velvet seat, one can read the names of a hundred and three Montparnos—some of whom one used to eat and drink with, laugh and quarrel with—cut like a decorative frieze on the stone-colored walls, in an alphabetical listing, beginning with Apollinaire and Aragon and ending with Vlaminck and some mysterious unknown whose name begins with "Z." On the list of those who frequented the Café Rotonde and have since known fame are Cocteau, Radiguet, Modigliani, Max Jacob, Foujita, Picasso, Braque, Honegger, Hemingway, Henry Miller, Bromfield, Utrillo, and, in a class by themselves, Lenin and Trotsky, who used to sip their *cafés-crèmes* there in 1915, when in exile. The lights go down, and a Russian version of Dostoevski's "The Idiot" comes on the screen. One can smoke in La Rotonde—the only small Paris movie that permits this—and while you hold a cigarette in your left hand, with the other hand you adjust a simultaneous-translation speaker beside your right ear. It is a metal disc mounted on a flexible stem, and in the present case it gives you Joseph Kessel's French translation of the Russian screenplay, recited by a French cast and so precisely correlated to the lip movements of the Russian actors on the screen that if you have finished your cigarette and clap your left hand over your left ear, shutting out the Slav sounds, you believe, wide-eyed, that the Russians are speaking French. Ocularly, the film looks more like Dumas than Dostoevski. Directed by Ivan Pyriev and covering only the first half of the novel, it is a big spectacle in color, with snowy Moscow street scenes as beautiful as if done in Hollywood. Its importance is that it inaugurates the Rotonde's policy of showing intellectual foreign fare worthy of the international intelligentsia that made the café famous.

August 12

The French Assembly probably deserves a word of mention right now, if only to remark that it has recessed for the summer, with its departure arousing even less public interest here than its presence did. It had been sitting since late spring, for three of the five months it is allowed to sit under the Fifth Republic—which is "only faintly parliamentarian," as a political wit recently said, in disapproval. The Senate has closed, too. Made up of older, tougher politicians, many of them reëlected as popular leftovers from the Fourth Republic, and with nothing like the Assembly's majority of débutants and members of the neo-Gaullist party (which seems increasingly to mean Gaullists who do not agree with General de Gaulle), the Senate in its recent Luxembourg Palace session at least showed signs of life by occasionally kicking with old republican vigor against the present government's traces.

More than a word must be said about the latest state of affairs surrounding the Algerian war, which never stops. General de Gaulle reportedly described it a fortnight ago as "this senseless war," adding, "Peace is a necessity. This absurd war concerns us all, and you must help me to end it." He is supposed to have said this off the record to the colored officers—largely Negro ex-colonials, mostly from Africa—newly elected to the Senate of the French *Communauté,* de Gaulle's Fifth Republic extension of France's vision of the future. His summing up of the war as *"insensée"* and *"absurde"* is simply a new reflection of the dichotomy of opinion on how to halt the struggle in Algeria and settle the country's future political status that has existed from the first between the General and his hand-picked Premier, Michel Debré. In Debré's most recent statement about the war, made before the largely white French Parliament here, he seemed to speak of France as being personified by himself, rather than by President de Gaulle, when he declared, "France would do anything anything at all—to keep Algeria French." While the biggest French military effort of the entire five-year Algerian war, called Operation Binoculars, has been going on for the last two weeks in the mountains of Kabylia, de Gaulle has been exploring tentative solutions with Algeria's neighboring Arab countries, as his own way of discovering a means of ending the bloodshed. It is known that the King of Morocco, in the interests of peace, is willing to undertake a mission for the exchange of information between the Arabs and the French, though if he is to furnish his French interlocutor (General de Gaulle) with information on the state of mind of the Algerian rebel army chiefs and of the leading Ministers of Le Gouvernement Provisoire de la République Algérienne in exile, whom he recently received at Rabat, he will presumably have to know the real intentions of General de Gaulle in regard to Algeria, which the General's ambivalent statements have so far not made clear to His Majesty (or to millions of French). For this purpose, a confidential meeting with de Gaulle was arranged on the King's recent arrival here for what was elegantly called "an ablation of his tonsils." But before the two could meet, the King had to return hastily to Rabat to mend Morocco's political and financial fences; the tonsillectomy was performed there, too, and he has been recuperating in the Royal Palace ever since.

Added propaganda tension among the non-white races came last week from the Pan-African Conference in Monrovia, Liberia, where the nine independent African states, instead of concentrating exclusively on their own troubles, unanimously voted that "France should recognize the Algerian people's right to independence, should withdraw the French Army, and should start negotiations," which was admitted here to be a diplomatic victory for the Algerian rebel cause. But there was also a big diplomatic victory for de Gaulle reported from Monrovia in the conferees' statement that "his prestige has remained intact among all the delegations." However, a real international crisis on the Algerian question is now considered possible in New York at the mid-September session of the United Nations General Assembly, when the necessary two-thirds majority of the delegates may vote that France should give Algeria her independence.

The one bright aspect of the Algerian problem is that during this past fortnight Premier Debré, who at least sees eye to eye with his President on the industrialization needed to raise Algeria from the status of an undeveloped area, visited there and brought back to his Paris press conference an extremely impressive report on the opening and hitherto unmentioned results of de Gaulle's Constantine Plan—the economic new deal for Algeria that in the General's view is the costly but unavoidable evolutionary accompaniment to France's "military pacification," as the Algerian war is officially still called here. Since May, the Fifth Republic's new Société Algérienne de Développement et Expansion has approved seventy-two applications from firms willing to invest about thirty million dollars in the country (with preferential tax relief), thus creating six thousand Algerian jobs at the outset; a big steel plant is going up in Bône; and the pipeline from the rich Hassi Messaoud Sahara oil wells to the coast is almost completed. It is this fabulous oil deposit on which the French state is gambling to pay for the Constantine Plan. Sixty-two thousand new jobs for Algerians are expected to be available by this time next year. Also by this year's end, private industry will probably have furnished seventy-eight million dollars more in investment money—which takes spunk, pioneering spirit, and no doubt patriotism, too, considering the political risks that even optimistic believers in L'Algérie Française cannot close their eyes to. Peace in Algeria is still the acme of the hopes of millions of French people under de Gaulle. But in the meantime it is de Gaulle's Constantine Plan that has apparently begun the real, secondary miracle.

August 25

This year, Paris has visibly built itself into the middle of the twentieth century at last. Whether you like it or not, you can see the proofs in the suburban ring of new apartment houses, containing three hundred thousand new flats, that now circles Paris where there were bosky vacant lots, or even green fields, a year ago. These *cité-parcs*—new inventions springing up all over France and given this new name—are architecturally identical and look monotonously like dominoes or dice, being of white cement spotted by rows of fenestration. Some have small, slab-fronted balconies (for parking the baby carriage) that are invariably painted red, blue, green, or yellow—a single color to a floor. One development, which claims to be the longest block of flats in all France, is just outside the northern Paris suburb of Pantin. It is a novelty in that it is built in the form of a wriggling snake; is nearly a half mile long, containing four hundred and fifty flats to house fifteen hundred tenants; has four acres of playground and lawn for children and baby-sitting mothers; and, by its pleasing serpentine pattern, gives light and air to all. Huge square fifteen-story towers are going up at Boulogne-Billancourt, near the Renault auto works; new five-story apartment houses have appeared in the forests near the Seine at Vernouillet, Rueil, Argenteuil, and dozens of other suburbs; and in Paris itself the Marché aux Puces, the old flea fair at the Porte de Clignancourt, is, alas, going to be more decently lodged nearby to make room for domino-flat buildings. Three miles outside the suburban town of Poissy, seat of the Simca auto works, in what has been for thirty years beautiful blue-green fields of cauliflowers, there now rises, like a strange urban excrescence, a *cité-parc* of forty-five apartment buildings, with twenty-two hundred flats housing eight thousand people, and featuring three twelve-story skyscrapers alien to such an empty sky.

Yet even at this rate of construction, France's housing shortage could continue for another century, according to statistics. The rent for a modest flat in these new buildings around Paris—three rooms plus kitchen, a *salle d'eaux,* with hand basin and shower, and a separate lavatory—runs to about fourteen thousand francs (twenty-eight dollars) a month; if a workman wants to buy a flat in a new coöperative building, he must figure about a million and a quarter francs (twenty-five hundred dollars) per room. Previously, the economical French used to figure on spending from four to eight per cent of their budgets on rent—living without modern comforts but saving money. (Even today, forty per cent of French *logements* lack running water, the water being down in the court; seventy per cent lack private lavatories and ninety per cent lack private showers or baths.) Only the desperate postwar housing shortage could have driven today's families to spend what amounts to twenty per cent of their budgets on rents that include modern conveniences whether they want them or not. Over the centuries, French fastidiousness has always leaned toward good food, on which a third of what a family earns is still spent. Nobody saves money any more; it is eaten instead.

This eruption of new housing is entirely owing to de Gaulle, who ordered the 1959 government building credits jumped by thirty-five billion francs, to a total of a hundred and ninety billion; created six new building societies under government control; and has favored what is called the Habitation à Loyer Modéré, or H.L.M., plan for the working class. There comes a moment in the modern life of a magnificent, previously regal, highly stylized, great old city, whose architecture has made and been part of civilized history, when breaking one's heart over the jerry-built monotonous ugliness of H.L.M. flats turns into a cardiac form of aesthetic hypochondria.

Another modern manifestation hereabouts has been France's first experience (which came only this month, showing how lucky it has been) with juvenile-delinquent gangs. Apparently, they call themselves Les Gadjos, though nobody knows the etymological derivation of the word, or has dared to ask. The peaceful citizens they attack with bicycle chains on the streets after dark have given them the name of Les Blousons Noirs, since black imitation-leather jackets are their tribal regalia. They are the French version of London's Teddy boys, of Germany's so-called "half-salted ones," and of Russia's hooligans—all based on the genuine, original New York "West Side Story" article. Les Blousons Noirs are also devotees of the late James Dean. In the Paris suburb of Drancy, where Paris Jews were first put in concentration camps under the Nazi Occupation, a hundred and seventy Black Jackets, under the orders of a fourteen-year-old chief, were arrested a fortnight ago; near Arras, a seventeen-year-old neophyte vainly tried to blackmail an obstinate old farmer out of half a million francs by threatening to set his wheat fields on fire; up in Finistère, that pious Breton land, sub-adolescent Black Jackets, aged only ten to thirteen, sacked their local school, about to open; on the beach at Les Sables d'Olonne, little Black Jacketeers almost disarmingly attacked a beach shop to steal money, balloons, and candy; near the Pas-de-Calais, a thirteen-year-old beat up an old gentleman in an odd attempt to force him into adopting him legally. In the Parisian Thirteenth Arrondissement, forty Black Jacket boys staged a real coup in the Rue Brillat-Savarin—first in warfare against another gang, then against the local post office, then against all passersby. ("If we didn't like their faces, we smashed them in.") Their parents, called to the police station, could not believe their eyes when they saw their sons there. The Black Jackets made their first national appearance in Cannes early in August, when forty of them fought it out with the local police, terrifying holidaymakers in décolletage. They act as they do, or so they say, *"parce que nous nous ennuyons"*—"because we are bored." It is France's latest *mal de siècle.*

1960

March 23

It is now painfully evident that de Gaulle's troubles, accumulating increasingly in the past month, throughout the last fortnight, and even over the weekend just gone by, have seriously affected the temper of the French nation and how it feels toward him. His once radiant popularity has begun to fade, and the docile obedience of Parliament has dangerously diminished. For nearly two years, he has been the extraordinary, salvational leader, who came to power as a symbol of new hopes and faiths for the French and for France, and who even appeared to be saving French republicanism by using his high hand to perform the labors that Parliament had demonstrably and repeatedly failed to be equal to, or even to have time for, because of its constant, preoccupying power quarrels, its rivalries, and its tattered patriotism, which left a vast rent in the fabric of the state, through which successive governments fell to the ground, like lost valuables. The drama of this change in attitude toward him is dated precisely, like a historical event. It began on January 24th—naturally, in the purlieus of Algiers—when the French Army clique of *ultras* again defied him, and were punished by him for mutiny against what seemed his hope of soon ending their war with the rebel Algerian forces by means of a negotiated cease-fire, to be backed by some far-future vote on Algeria's self-determination or, at worst, its independence. At least, this had seemed de Gaulle's hope from what he had previously said, even when he did not say it clearly. Unfortunately, the French people's overstimulated devotion to him and to peace began expiring under the sudden illogical contradictions that inexplicably came flowing from the General's highly educated mind. He declared, in a *volte-face,* that the war must go on until the Algerian supporters of independence were utterly destroyed—and with that the French Army again triumphed in an old-style colonial military victory, climbing back into the saddle of political power from which he had just thrown it. The most painfully popular cartoon in Paris was one by Vicky, reprinted from the London *Evening Standard*. It showed de Gaulle standing on his head in war-torn Algeria while stating, "As I was saying last week . . ." De Gaulle came to power two years ago largely on his implicit promise that he would end the war, now proceeding apace through its fifth year. Opinion here seems to be that he only recently found out that he could not stop it—his tragedy and France's. Until he made this discovery, his hopes led him to speak ambivalently, the hopes and facts in the case diverging importantly. To the French now, the only clear fact is that there is to be no peace. Another opinion, more special, is that General de Gaulle, because of his mystique, his professional career at arms, and his intimacy with French history, still believes that only the Army symbolizes the unity of France, and is psychologically unable now to separate Army and country. This, at any rate, is the bare, condensed scenario of the suspense drama currently going on here, of which the last act is not yet written, and perhaps not even imagined.

General de Gaulle's civilian troubles with the parliamentary deputies started only a fortnight ago with a demand by a legal majority of them—as fixed by his own constitution—that he convoke the Chamber in a special session before its normal April term in order to talk over the financial crisis of the French farmers, out on a limb and left behind by the country's economic recovery. This, some newspapers think, was parliamentary Machiavellism to flatter the peasants and at the same time to use the poor price of Brittany potatoes as a political cover for the unbearable national cost of French blood and money being spent in Algeria, the deputies' real aim being to weaken him by adding to his unpopularity on any pretext. In this the General helped them by refusing to grant their demand, quoting multiple reasons that he said were also contained in his constitution. This refusal has, in turn, raised the graver question of whether de Gaulle, the man of acknowledged probity, the leader with a fetish for *la légalité* such as French kings claimed for ruling, has used his constitution in a constitutionally illegal way. In any case, his categorical, courteous refusal to give the deputies what most citizens thought they had a right to demand has had a spirited and bad national reaction—quarrels among jurists, millions of words of argument and criticism in the press, increased public bewilderment, and even hostility. In a magazine article, Guy Mollet, the Socialist chief who helped put de Gaulle in power and has supported his regime, has warned him, "No further mistakes can be permitted, *mon général!*" The phrase "the Sixth Republic" has been spoken and printed (so far sparingly), as if the Fifth Republic might be brought down by the exceptional Frenchman who founded it.

April 12

It is too bad that President de Gaulle's welcome home included grave trouble with the French farmers. They are now a changed lot, know it, and are at the end of their patience. Even just after the war they were still picturesque, were stilled called *les paysans,* were still supposed to be rich enough to stuff their straw mattresses with bank notes, still plowed their fields with Norman stallions or oxen, and were the politicians' pets, constituting twenty per cent of the population and a huge vote. Now they call themselves *les cultivateurs,* wear yellow gum boots *à l'américaine,* are backwardly trying to industrialize their production, are a dwindling class because their sons go to work in town for real money (and for plumbing and movies), plow with tractors, use chemical fertilizers, are in debt, and are unpampered by the politicians, who, they have just said, treat them like "poor relations." In France's contemporary social family, this they certainly are. They have now pushed their financial crisis over the last two years to the front, where it belongs as the nation's leading domestic problem. A week ago, close to half a million *cultivateurs* held mass protest meetings in seventeen of France's biggest market towns. The farmers' union—La Fédération Nationale des Syndicats d'Exploitants Agricoles—ordered them not to take along their pitchforks, the age-old weapon and symbol here of peasants in revolt for their rights. The union's preliminary mass meeting in Amiens in February had ended in a riot with the state police, in which tear-gas bombs, pitchforks, and paving stones were used, and hundreds were injured. At last week's monster meetings there was less violence but more organized indignation, with slogans on placards that showed which way the wind is blowing. At Nancy, the slogan was "Better Be Rebellious Than Resigned"; at Tours, "We Don't Want to Be Slaves of the Fifth Republic"; at Nantes, "The Peasants Are Certainly Worth As Much As One Atomic Bomb, *Mon Général.*" Considering the French farmers' addiction to *"la goutte"*—the "drop" of hard liquor, like Calvados or *marc,* that they lace their coffee with and swig by the glassful whenever they get together in the *bistros*—the meetings were pretty orderly, instructive, and constructive. The main farmer speaker at the Laval town meeting spoke for all ordinary small French farmers (big farmers with two-hundred-acre farms being rich and rare) when he said, "This is our last warning to the government. It knows our situation. We hold it responsible."

Eh bien, so far the government's proposed reform bills offer too little too late, in long-term plans for 1961–63, which are of no earthly help to the peasants' tight budgets right now. The deputies will certainly try to speed things up when Parliament opens in two weeks. The projects are for rural equipment, more agricultural schools, livestock inoculations, water supplies, and sick benefits for all farm workers (something they don't get now), but there is no project for the farmers' principal demand, which has been ducked by the government for fear of renewed inflation—guaranteed adequate farm prices tied to rising living costs, as they used to be. It is the French cities that have created and enjoyed the recent big prosperity boom, the advancing technology of France, and the high wages of workers, who can pay for and eat like the bourgeois, with the same Sunday *filet de bœuf* and daily *bifteck*. French farmers, who are nature's conservatives—the perfect producers of perfect rich-tasting produce by nineteenth century methods, on farms still as lovely-looking as landscape gardens—are way behind the times in living standards and are outraged that factory workers can buy the best they raise when they themselves cannot afford to buy even the average of whatever the factory workers manufacture. What drives farmers really mad is to read in the local-newspaper produce quotations that the Paris housewife pays three hundred francs for a kilo of leeks when the farmer has sold four kilos for half that—the difference going largely to the Halles' great central-market middlemen, France's most hated locusts. Farmers here want land mortgages at three per cent interest, like their neighbors in other countries, instead of at the twelve per cent the French pay, and want government money loans at one per cent, instead of three and four per cent, which is unheard of beyond their borders. Briefly, they want to get out of debt; want to modernize, because government technocrats assure them that it pays; and, most of all, simply want to make money again. France cooks and passionately enjoys eating the best food in the world. Maybe she had better help her fine food raisers to prosper.

April 29

On Tuesday, the first regular 1960 session of the French Parliament, in recess since last December, opened with the expected lukewarm Socialist and Radical threat to try to overthrow the government of President de Gaulle's Premier, M. Michel Debré, by a motion of censure. The censure motion against the government (for, by President de Gaulle's Constitution, de Gaulle is not responsible to Parliament, may not himself be censured, and, indeed, may be charged only with treason) is largely a political gesture for the record, to show that the Socialists and the Radicals, leaders of the Left Opposition, now definitely existing, deeply disapprove of the General's refusal to call a special session of Parliament last month to discuss the agricultural crisis, even though the Opposition had obtained the legal right to demand a session by assembling a constitutional majority of deputies—a double brushing aside that both Parliament and the farmers are still smarting under. Naturally, Debré's opening speech to Parliament on Tuesday was on the agricultural problem. In it he first recalled the classic agrarian formula that the two breasts of France are her tilled land and her pastures—the bread and milk to feed her people—and then went on, less bucolically, to matters of hard cash, calling the farmers' demand that their prices be automatically hitched to rising living costs "a one-way infernal machine" that would inevitably make all prices run uphill.

This being the first significant anti-government political maneuver against the Fifth Republic, comparisons with the Fourth Republic, which used to mow governments down like weeds, are now being freely made, if only as a means of estimating the Fifth's strength. Everyone agrees that any government under the Fourth would have fallen flat by now but that it would have been too smart to let the farmers' predicament fester. Another difficulty for the government will come from this week's ousting of M. Jacques Soustelle from the Gaullist political party, the Union for the New Republic, or U.N.R., which, ironically, Soustelle founded and is now splitting. As a champion of Algeria's integration with France and of war to the finish to exterminate the rebels, he will now be able to raise an uncontrollably loud Opposition voice among the *ultras* on what still remains the vital emotional and political touchstone of France—the Algerian problem and lack of peace.

June 22

Monday's news of the sudden acceptance by Le Gouvernement Provisoire de la République Algérienne of President de Gaulle's offer, in his last Tuesday-night radio speech (today's decibel-laden substitute for former secret diplomacy), of plans for a talk here in Paris "to find an honorable end to the war that still drags on" roused only a mixed measure of hope and dubiety in the French people. It was as if the over-familiar alloy of doubt had dulled the shining radiance of hope, and its value. Four years ago, the Resident Minister of Algeria, M. Robert Lacoste, had assured the French that the war was in its "last quarter of an hour"—that idiotic, ill-computed phrase being his only memorable contribution to the war. The Socialist Mollet government, brought to power on its pacifist intentions, merely waged the war all the harder after M. Mollet himself was pelted by the natives with tomatoes, in season at the time of his friendly opening visit to Algiers. Even General de Gaulle, from whose exalted superiority something as high as a miracle was expected, founded his Fifth Republic on his belief, shared by the majority of the French people, that he could bring an end to the war. Now, two years later, he seems to have brought it within sight. There is still a large gap in vision, however, as was mentioned in the blunt communiqué of acceptance issued by President Ferhat Abbas, of the so-called Algerian Republican Government. In it, Ferhat Abbas said forthrightly that it is his government's belief that if a "sincerely free" referendum on self-determination were actually offered by the French to the Algerian people, "their choice, without doubt, would be for independence." This probability has been met head-on only by the serious, independent paper *Le Monde,* in an editorial of the kind that is written exclusively at grave moments in French affairs by its austere editor, M. Hubert Beuve-Méry, under the mystifying nom de plume of Sirius, as if the Dog Star were suddenly shedding needed extra light. In the Monday-night front-page Sirius editorial, entitled "Hope," M. Beuve-Méry boldly said, as did no other editorialist, "It is alleged that any accord made with the Army of the Algerian National Liberation Front will open the way to independence sooner or later. This is true. But when all of Africa is shaking off its European tutelage, can it be believed that Algeria alone will be able to remain outside this gigantic revolution, and that a million French can impose their law indefinitely on nine million Moslems in full demographic upsurge?"

The answer is "yes"—at least, as given that same Monday afternoon by what was pedantically called "Le Colloque de Vincennes," a symposium of ultra-Right Wing politicians (mostly mavericks who were once in more moderate parties) held in the suburb of Vincennes, where they declared themselves a possible new political party against de Gaulle, and certainly against Algeria's being regarded as anything but French forever. Their basic fear—and they are not alone in feeling it—is that an independent Algeria might all too easily change from a formerly rebellious French colony into an obedient Kremlin colony, to judge by Moscow's and Peking's recent violent propaganda for Algeria's liberation. The Vincennes colloquy had long been planned, but the sudden news of the impending de Gaulle-Ferhat Abbas talk so increased its timeliness that it was attended by thirty-five deputies, some senators, some former Premiers, and two hundred important figures from universities, big business, labor unions, and government administration, along with a crowd of political experts on the *qui vive*. The principal speakers were M. Jacques Soustelle, former Gaullist leader and now the most dangerous, brainy anti-Gaullist leader, and M. Georges Bidault, once the sanctified chief of the liberal Catholic M.R.P. party, who, after de Gaulle's Tuesday radio speech, declared that de Gaulle's proposal to Algeria was "sombre folly" and possibly tantamount to treason—which, even in France, was an almost incredible attempt at verbal political assassination. How violent may be the reaction to de Gaulle's peace policy by the French Army *ultras* in Algeria itself nobody here is yet sure of, except to recall the bloody example given on January 24th, when they barricaded the center of Algiers in their mutinous maneuver against the Paris government—a mutiny that President de Gaulle, suddenly becoming General de

Gaulle, broke by his military orders. It is hoped here, and fervently, that French political civil war does not break out again in France's Army.

Among the scars suffered by the French social fabric through the five and a half years of this peculiar war have been the public's enforced awareness that torture has reportedly been practiced by elements of the French forces in Algeria, on both French and native sympathizers to the Algerian rebel cause, and that reputable French papers carrying carefully authenticated reports of such shocking things have been seized by the police, apparently on what amounted to Army orders. These practices have constituted a double offense to French republicanism in the Fifth Republic—first in the disgraceful, anguishing abuse of human bodies, and second in the violation of the freedom of the press. On June 3rd, *Le Monde* reported that its June 2nd edition had been seized in Algiers for having published an article by Mme. Simone de Beauvoir about an Algerian girl, Djamila Boupacha, who, as the examining prison gynecologist attested, was indeed suffering acutely, just as she maintained, and who claimed that she had been vilely tortured into making a false statement about participation in a bomb-throwing incident in a café—the Army's only counter-claim being that she had made her admission. Her awful story, which was first talked of in March, drew a great deal of shocked comment here. Last week, she came up for trial in Algiers before an Army tribunal, whose judge ordered the proceedings postponed, "pending further information," until July, when she can be condemned to death. These strange violations of civilized republicanism are considered by the French to be the humiliating mysteries of General de Gaulle's Fifth Republic and of his portion of the five and a half years of the demoralizing Algerian war—*la sale guerre,* the dirty war, as the younger French generation calls it—which he now may be bringing to an end, without victory.

Last Saturday, June 18th, was the twentieth anniversary of de Gaulle's first intimate and influential historical relationship with his *belle France,* which he has, on his heights, continued whenever possible. His first words to her were those he spoke from the B.B.C. in London, dressed in the uniform of a defeated army but wearing white gloves. This was not that prophetic, legendary appeal that we have all come to think it was—the one beginning "France has lost a battle; but France has not lost the war!," which was printed on posters and pasted on London's walls a few weeks later. The appeal he made that June day began, characteristically, with identification —"I, General de Gaulle"—invited the remains of the French Army in England to get in touch with him, and ended, "Whatever happens, the flame of French resistance will not, must not, be extinguished." Last Friday evening, he lit the first vigil flame before the enormous new Resistance Memorial that has been constructed below the old suburban fortress of Mont-Valérien. It is the only monument to the heroic dead of this last war that France, in all conscience, could erect—honoring the dead Resistance fighters, the dead voluntary patriots who fell for Free France, and, in particular, the forty-five hundred Resistance fighters whom the Germans captured and imprisoned in Mont-Valérien, where they were shot by Nazi firing squads, many of them just before the Liberation. For this reason, Mont-Valérien has become the sacred site of the French Resistance. Before a huge crowd of fifteen thousand French, drawn by memory, sorrow, or national and personal pride, and composed of former Resistance members, men and women from all over France, and of the families of those who failed to survive, there was a military torchlight procession that accompanied the placing within the crypt of sixteen coffins, containing nameless dead, which have been sepulchred in the fort ever since the Liberation. The Versailles Cathedral choir sang "Le Chant des Partisans," that melancholy Resistance song, and at midnight twelve cannons sounded in salute for the dead. The next day, the public came by the thousands again, to view the memorial by daylight. It is handsome and dramatic, with noble, solemn proportions. Below an old wall of the fort on its hill has been built an immense lower wall, three hundred feet long, of a reddish, almost flesh-colored, stone, dominated in the center by a gigantic stone Croix de Lorraine, nearly forty feet high. On either side, ornamenting the wall, are eight bronze *haut-reliefs,* each by a different French artist and each depicting some phase of the Free French struggle. One, of twisted hands and barbed wire, symbolizes the concentration camps; another is a huge human heart in an opened breast, symbolizing torture; another is a crane flying with wings drooping, like a parachute; a wounded lion struggling with a Hydra-headed beast stands for the desert fighting; and for the seamen there is a youth fighting with an octopus whose twisted tentacles vaguely suggest the Nazi swastika. The memorial is officially called Le Mémorial National de la France Combatante.

Those who fought these battles were members of de Gaulle's private army, and the new touching and magnificent structure is their memorial. But, in its way, it is de Gaulle's monument.

July 6

Until Monday afternoon, when the announcement was made of the Algerian rebels' spirited and profoundly disappointing refusal to send another cease-fire delegation to France, because it would not be "opportune in the circumstances," millions of French had thought that the Algerians' earlier acceptance of President de Gaulle's June 14th offer to discuss "an honorable end to the war" looked like the best chance for peace in the sixty-eight months since the fighting began. For five days last week, these optimists were able to let their hopes play freely while the first fact-finding Algerian delegation, to its surprise, sat housebound, invisible and incommunicado, in the handsome old Prefecture building at Melun, forty-five kilometres outside Paris. Thus, the opening difficult motions for honorably ending the war got off to an awkward start, at least in the view of the influential chief Algerian emissary, Maître Ahmed Boumendjel, a genial, popular former Paris lawyer who speaks French better than Arabic (doubtless one of the reasons he was chosen), who only lately joined the rebels in exile, and who, with his aide, M. Mohammed Ben Yahia, *chef de cabinet* of President Ferhat Abbas, found himself confined in Melun for strictly secret conversations with the two French delegates to the cease-fire conference—one the General Secretary of Algerian Affairs and the other an elderly high Army officer. Nor was any information as to what the four were really talking about given to the press to hand on to the impatient and deeply interested French public. The journalists who went to Melun were kept outside the grilled fence that shuts off the Prefecture's courtyard just as strictly as the Algerians were kept invisible, day and night, inside it—the last thing they had expected or desired in a meeting of such vital Franco-Arab importance, from which they wanted their libertarian point of view to be heard around the globe.

As French citizens themselves have repeatedly noted with exasperation, official public relations are peculiarly unbalanced in this present republic, whose chief not infrequently communicates with his compatriots and the world in extensive, splendid literary addresses that are ornaments to the French language. But his government, supposedly on his say-so, is a great one for keeping mum, especially in dispensing information that might do it or the public some good. Of the brief Melun meetings (at which the two French representatives never shook hands with the Algerians—strange protocol indeed), a disgruntled foreign observer noted, with obvious restraint, "The past week has shown French diplomacy and treatment of public opinion at its most baffling. The Frenchman in the street, on whom the regime must count to resist political extremists, and whose own fate is closely involved, has had to go through an incredible game of trying to guess what half-hints, contradictory statements, and leaks from Tunis meant or did not mean." Even the government's all-important communiqué at the end of the five restricted days at Melun merely said that conditions for negotiations on the honorable end of the war had been laid down, without giving any reader, Christian or Moslem, the slightest idea of what those conditions were. The last curt line of the communiqué was, however, only too comprehensible. It said, "These preliminary conversations having been concluded, the emissaries are now to return immediately to Tunis" (where the Algerian rebel government sits out its exile)—a phrase that sounded to some alert-eared Parisians like a sharp dismissal indeed of the emissaries, and perhaps also of any immediate possibility of Algerian peace.

It is probably unique, even in ultra-modern state documents, for a complaint about not being allowed to use the telephone to figure in an official paper issued by a people at war, with peace at stake. Yet in the Algerian Provisional Government's Monday list of protests against the remarkably protocolar restrictions laid down by the French for the conduct of any future Algerian emissaries (one of whom might be Ferhat Abbas himself), there is mention of the fact that such emissaries would be forbidden even "to communicate indirectly, by telephone," with certain captured Algerian Ministers held here either in prison or in military fortresses. Indeed, they would not be able to telephone anybody, since no one of the Algerian delegation would be permitted to make contact with "anyone, in any manner, either in France or outside France, except in Tunis." They would also be forbidden to have any communication with the press; would be obliged to live in total isolation; and so on and so on. However, in all that list, it is the unexpected, intrusive image of the telephone in relation to that savage, primitive, empty North African countryside where the Algerians have been fighting their war of rebellion that sticks like a Dali landscape in one's mind. Both the Algerians and the French must now face the fact that the entry of peace in their relations has been proposed, that both their peoples desperately want peace to come in, and that neither the rebels nor the French can afford to risk being criticized by world opinion, or even by their own nation, for having been the ones responsible for slamming the door against peace, which, logically, would mean going on with the useless war.

The sudden pressure of French public opinion on the welcome possibility that peace will indeed come to pass has been building up here for a month, and reached a climax this last week. In what way, or on whom, this organized energy will now expend itself after the Melun Algerian impasse remains to be seen, and might prove to be a public problem, as some of the French already fear. On Thursday of last week, a remarkable declaration was signed jointly by four powerful French unions. These were the Communist-dominated Confédération Générale du Travail; the Confédération des Travailleurs Chrétiens, which is the Catholic workers' union; the Fédération de l'Education Nationale, the teachers' union; and the Union Nationale des Etudiants de France, composed of ninety thousand university students—the most influential aggregation of well-educated youth in all France. The declaration said, in part, that "at this moment, when the Algerian drama is entering a decisive phase . . . the above organizations affirm in unison their desire to see negotiations really opened and carried through to their normal conclusion—a cease-fire and an agreement, with indispensable guarantees, on putting self-determination [of the Algerian people] into motion. These organizations have together renewed their resolution to use all the means at their disposal, including that of the general strike, as an answer to any insurrection or *coup d'état* that might tend to impede the Algerian peace, or even tear down further the essential democratic liberties [of France]." This is, of course, a double warning—to the de Gaulle government to get on with ending the war, and to the *ultras* not to repeat their bloody *coup d'état* of January 24th in Algiers, raised in protest against de Gaulle's increasingly liberal Algerian policies.

To get back to the students' union, which is extremely important, it has already paid dearly for its energetic stand on peace. On June 6th, its officers resumed the union's old relations with the Union

Générale des Etudiants Musulmans Algériens, which in 1958 had been declared officially dissolved, and therefore illegal, by the last government of the Fourth Republic, because of the Moslem students' loyalty to their home cause. Upon this renewal of relations, the two student groups, French and native, signed a communiqué setting forth their conviction that "discussion between the French and Algerian nationals is possible, and the only competent means of bringing the colonial war to a close." On June 16th, in punishment for its association with the Moslem student union, the French government's Minister of National Education cut off the French student union's annual subsidy, which it had been receiving along with other educational or cultural youth movements, and which would have amounted this year to eighty thousand New Francs, in part payment for its services in organizing student housing, distributing scholarship monies, and doing other university jobs. On Sunday, June 19th, the student union held a Paris meeting to which a hundred and fifty or more regional delegates came from all over France to declare whether they still backed the peace stand, earlier voted in a Lyon meeting by a massive majority. The delegate of the Ecole des Langues Orientales, founded in Paris by Louis XIV, thought the *rapprochement* with the Algerian students questionable. The delegate from Paris-Science also thought that the student union was on the wrong track. But Paris-Pharmacie and Paris-Médecine delegates thought that the Minister of National Education's removal of their subsidy was an attack on students' rights and liberty of thought. Motions criticizing the union's officers for their realliance with the Moslem students and demanding that the union return to its previous non-political stand on current national events were both heavily defeated. As the union's president—twenty years old and a Sorbonne man—is reported to have said, "Apoliticism is impossible today. The Algerian war is too present all around us. We don't want any more of this war. Our increased political substance is what will unite us against it."

Tuesday of last week had been chosen as the day for demonstrations all over France in favor of negotiating the peace—demonstrations that the government forbade, fearing public disorders while it was itself supposedly negotiating at Melun for the opening of peace. In the provinces, at St.-Etienne, an industrial center of machine-making and ribbon-weaving, a crowd sang the "Marseillaise," hung a floral wreath on the statue of Liberty, and shouted

"Vive le succès de la négotiation!" before being dispersed by the police. At Roanne, another important manufacturing center, the crowd shouted not only for peace but against the government's ruling against their shouting for it. At Grenoble, from among a large crowd of demonstrators, eight citizens calling for peace were arrested, including a priest and a lawyer. At Poitiers, the police used their clubs against the crowd, and at Nantes they used tear gas. At Châlons-sur-Marne, the state police were called in from nearby Reims to deal with the peace manifestation. At Lille, Brest, Le Havre, Nîmes, Dijon—some of the best-known tourist towns—the police were also forced to break up the congregations of peace demonstrators.

For the past twenty years, June has been a month of painful historical recollections for the French. And over that fifth of a century they have suffered almost continuous small, troubling wars. But the endless Algerian war, which the young French angrily call the "Hundred Years' War," has been the most destructive of the essence of France itself. Certainly it had been hoped that by this June's end the seemingly century-long war might at last be coming to its close. Not in years has there been such active peace sentiment in France as in this deceptive past month of June.

July 19

"Stevenson, the eternal 'possible' candidate for President if the other candidates first eat each other up . . ." This was the astonishing manifestation of pro-Stevenson sentiment carried in a Paris headline just before the Los Angeles convention opened, Mr. Stevenson having been the French favorite for President on three occasions now, whether running or not, and even after he has lost. There was not a crack Paris journalist sent to the California Sports Arena who failed to star the prodigious emotional acclaim that greeted Stevenson's entry there—"slender, elegant, aristocratically bald, his eyes bluer than ever under the camera flashes and spotlights," *Le Monde's* special reporter cabled back home, which was accurate, surely, but dreamy. The convention's immediate choice of the youthful Senator Kennedy has led to long French analyses of him and of the recent acute dissatisfaction here with the "fatigued" President Ike's administration, which has automatically helped popularize Kennedy in France as "fresh and dynamic," with a "rather worrying virtuosity." Brought up to the practice of gerontocracy, the French earlier had a filial respect and affection for Eisenhower, which has given way this year to a list of criticisms of his Republican administration that clearly show the French hope of a change for the better in a Democratic victory, with, if must be, a very young President. "For eight years, the Americans chose 'the father' in politics," one expert has just pointed out. "Today, the American public sees the perils of it. The country's traditions—even its security—are at stake. The dramatic, dangerous events in Cuba, the aerial incidents over Russia, have added to the American public's confusion, giving them the indisputably mounting impression that they have been outclassed in the power struggle and in diplomacy, as well as in spying and sputniks." "Right now, it may be that youth has to be trusted," one Parisian was heard cautiously saying in a discussion of the coming election struggle. The strongly liberal Left Wing weekly *Express,* most influential journal of all among intellectuals and the young French, last week used a photograph of Kennedy on its cover, with the famous quotation on old men's wisdom from Hemingway's "A Farewell to Arms," translated into French: "No, that is the great fallacy; the wisdom of old men. They do not grow wise. They grow careful." Inside, *L'Express* used the recent pro-Kennedy piece by Mr. Joseph Alsop—also in French, of course. Kennedy's Harvard education, his ability to read important books, and his brain trust (quoted here as being composed of "M. Kenneth Galbraith, economics expert; M. Arthur Schlesinger, Jr., as specialist in social history; and Dr. Zbigniew Brzezinski, young chief of the Harvard group") make the French think that a Democratic White House could be a stimulating change. To the French, naturally, the most personally interesting facts about Kennedy are that he may be the first Catholic President of the United States and that he has already declared his intention of adhering to what the French liberals and the Left have for two hundred years called anticlericalism and, at the beginning of this century, obtained—noninterference in politics by the Church. But the greatest French hope in Kennedy is on the mundane side. They hope that with his youth, his sharp, educated brain, and his realistic eye he will institute a State Department that will look at modern geography as it really is, from Cuba to South America to Peking to Moscow and, above all, to Europe, with its intimate susceptibility nowadays to what Washington says, does, and thinks. Furthermore, most of the French certainly seem to hope that his Secretary of State will be M. Adlai Stevenson.

August 2

The so-called *"nouvelle-vague,"* or "new-wave," French films have increasingly aroused moral concern at high governmental levels, their libertinage having become a visible public fact in four of them that are still being shown here in Paris month after month. Since the de Gaulle regime's style is one of dignity and *bon ton,* its Ministers have been gravely embarrassed by the severe criticism these movies have received from the Church, family organizations, parents, and displeased adult spectators, and by such denunciations as that of the ultra-conservative Professor Pasteur Vallery-Radot, a well-known explosive member of the Académie Française, who said their aim was *"gangsterisme ou érotisme."* Questioned recently in Parliament about French films' immorality, the Minister of Information could only say that he planned more effective censorship. This is a government measure that French liberals have distrusted on principle ever since the notorious censoring of Baudelaire's "Les Fleurs du Mal" and Flaubert's "Madame Bovary," which survived to become classics, and which, of course, none of these movies, with their gifted or meretricious sensuality, faintly resemble. The Minister of Justice is working on a bill for pre-censorship, or censoring the scenario before the film can be made, with the logical idea of preventing very young adolescents from working as actors in any film—as two did in "Les Régates de San Francisco," one of the four new-wave pictures still showing here—that the censor will later mark "Forbidden to Anyone Under Eighteen." The ticket sellers for these *nouvelle-vague* films are strict about their job. They demand that every young film fan show his *carte scolaire,* which lists his birth date. The movie that gave the *nouvelles vagues* such a bad name (many have been absolutely excellent) was "Les Liaisons Dangereuses 1960," directed by Roger Vadim, first husband and discoverer of Brigitte Bardot. It was the last film, unfortunately, that Gérard Philipe made before his death, and it still serves as an unpleasant obituary to his fame. It is running here, full tilt, as the single new-wave film that the French government has refused to grant an export license to, because it would give France such a black eye. You are missing nothing—a shoddy, queasy modernization of the classic novel that was ironically written against licentious corruption in high society by one of Napoleon's officers, Choderlos de Laclos. In the Vadim version, which ludicrously takes place at the ski resort of Mégève, the famous wicked lovers, the Vicomte de Valmont and the Marquise de Merteuil, and merely a rich, fast married couple, as if to make more respectable their demoralization of innocent little Cécile, who already looks like a pretty sophisticated young starlet. To its credit, the Société des Gens de Lettres, indignant at what Vadim had done with the old book, forced the addition of the date "1960" to the title. Since to the public this seems to imply modern orgies, the change has made it super-popular.

Considering the long-drawn-out hullabaloo over the *nouvelle-vague* films and the French government's recent involvement, this might be a good moment to set down some facts. Started two years ago, they were at first called *"les films des jeunes,"* and their fundamental quality was that they were created by a new school of very young, unknown directors, who saw things with brand-new camera eyes, had to make their movies on a shoestring, and so developed a fresh, realistic, compact, and stimulating technique, using décors that cost nothing and were true backgrounds to young life—sidewalks, cheap cafés, and beds. Several of these young men had been sharp-penned movie critics on the influential *Cahiers du Cinéma;* all were sick of the costly, ripe films made by their elders. Their new-wave movies ("Les Tricheurs," about the young immoralists at the Sorbonne, was a typical early example), like all those that have followed, are part of today's intimate fight between the mature and the young on both sides of the Atlantic. What these young directors have shown best is what they know best—the anarchic, spasmodic, lawless, and rebellious lives of certain modern young French, intensely lived within their own cycle. In these films, the short cut to romance is sex, the stencil for beauty is the nudity of lovers. The roster—by name, age, rank, and major film—of the most important of these young directors runs like this: Alain Resnais, thirty-eight, considered the master of the new generation, "Hiroshima, Mon Amour"; François Truffaut, twenty-eight, leader of the *Cahiers du Cinéma* wing, "Les Quatre Cents Coups"; Louis Malle, twenty-eight, "Les Amants"; Claude Chabrol, twenty-nine, "Le Beau Serge" and "Les Bonnes Femmes," his latest, which shows the best of him and also the worst; Jean-Luc Godard, thirty, also from the

Cahiers du Cinéma, "À Bout de Souffle" ("Breathless"), which is called the chef-d'œuvre of the new school, is now the big hit in Paris, and is the best French film of any kind this year.

"Breathless" is a Champs-Elysées-sidewalk and bed film, perfectly directed, that features a Paris *type* who has killed a policeman while stealing a car, his regular profession. He is in love with one of those pretty American girls in slacks and sweaters whom you now see on the Champs selling the Paris *Herald Tribune,* with the name of the paper writ large across their bosoms (a role well played by the American Jean Seberg). A tart, and tired of him, she tips off the police, who shoot him down in the street as he is making a getaway. The big bed scene, when she comes home and finds him in it, as he has often been before, is a comedy passage at first, in which, while he keeps talking, he drapes the sheet over his head and nude torso until he talks her under the sheet, too—the sheet remaining like a tent over them in a very daring yet utterly discreet dénouement. The film's terrific drawing card is Jean-Paul Belmondo, whose ugly attractiveness is the latest emotional rage—a tiptop young actor who was thrown out by the Conservatoire d'Art Dramatique because he was so remarkably ugly, they said, that he could never appear on their stage with a woman in his arms, since the audience would laugh. This turns out to be utterly untrue.

November 2

For several reasons, none of them agreeable, this present week is being called *la semaine algérienne*—in Paris, at least. In the recent sudden stampeding of French public opinion toward its old, impatient vision of peace, the seven-year-old Algerian war problem has now become the dominant national concern for France; for the survival, possibly, of the Fifth Republic; and certainly for the verbal genius of President de Gaulle to solve and finally give his clear answer to. Tomorrow, Thursday, there opens at the Palais de Justice the trial that has already been dubbed the *procès des barricades,* which will intensify the emotions of the diminishing group of the extreme Rightists, who demand that Algeria must remain unindependent, unautonomous, and purely French. Chief among the sixteen civilian insurgents who will appear in the prisoner's box is M. Pierre Lagaillarde, deputy from Algiers and mainspring of the bloody insurrection of last January 24th, when the *ultras* set up their barricades in the center of that city in rebellion against the distant authority of the French government and, more locally, against General de Gaulle's then recent declaration for Algerian self-determination. In the melee, twenty-four French people were killed. Also on Thursday, M. Jacques Soustelle, leader of a brand-new anti-Gaullist Right Wing integrationist party, which he candidly calls the National Front for French Algeria, will be holding a surburban caucus with his best-known followers, mostly former high Fourth Republic politicians like Georges Bidault, Maurice Bourgès-Maunoury, and Robert Lacoste. In order to appreciate all these incredible confusions now going on here, it must be recalled that Soustelle was the leading brainy figure behind the other Algiers insurrection—the one of May 13, 1958—which put de Gaulle into power in the first place; was chief of the Gaullist Parliamentary party, the Union of the New Republic; and was also de Gaulle's appointee as Minister of Information. Then, last February, de Gaulle dropped him from his government as a malcontent, after which the U.N.R. ejected him into limbo, where he has found other anti-Gaullist companions.

To all intents, the week will end on Friday night, when President de Gaulle will address his uneasy country in a televised allocution. France is waiting to listen to him with critically acute hope and a certain preliminary sense of alarm, anxious to see how well, at this perhaps crucial moment, he will be able to do for himself in his accustomed literary outflow of echoing, magnificent, though not always communicative, language. Now, according to the French, is the time for the General to communicate—now, when pragmatic faith in him has started falling, even though he is still master of France, with its citizens, and even history, once again waiting to hear him speak. It is true that he has been talking briefly, off and on for the past few months, in the conscientious majestic appearances that he has made in the countryside and in towns all over the land—a bareheaded, aged patriot still offering the more intelligent glories of France to provincial, often humbly chauvinist crowds, who have cherished his words and his appeals for national greatness and strength as they listened under their umbrellas in the rain. *"Aidez-moi, Français!"* he demanded in an appeal he made in one Alpine town, and was loudly reassured by cheers—as, indeed, he was almost everywhere. Talking in the Midi city of Menton, and again pursuing the topic of how France should be run and by whom, he declared, "The control of France belongs to those who have been made responsible for it, and it thus belongs, *par excellence,* to me. I say this without any beating about the bush." Not since Louis XIV's declaration *"L'Etat, c'est moi,"* at which no one in those times dared smile, have the French as a nation smiled so broadly as at de Gaulle's paraphrase. He has now been nicknamed Général Moi in the weekly satiric papers. Since the Menton speech, *Le Canard Enchaîné,* which is always against any government, and especially de Gaulle's, has been running a political gossip column entitled "La Cour," written in an uproarious take-off on the grand style of the court memoirs of the Duc de Saint-Simon, which the *Canard* has illustrated with a wonderful caricature of de Gaulle dressed up like the King, with a nose the size of a Versailles chandelier.

De Gaulle was called to power two and a half years ago by what seemed destiny—plus an enormous majority in a referendum—as the one man in France who might perform the miracle of bringing his country back to something like its old portrait. He has made France once again rich, sound, prosperous, handsome. Only in one major way, as the French deplore, has the miracle failed to come off—in Moslem Algeria. Part of the loyalty and hope still felt for de Gaulle lies in the fact that now, as then, there is nobody else of his patriotic height visible on the French landscape. In fact, there is no one in sight at all, of any experienced quality, who has not already failed at what de Gaulle has not yet succeeded in completing. The only real fear here among the general public and the Parliament is that he might abandon them under criticism and disappointment, and retire again to Colombey-les-Deux-Eglises and his solitary study of history, leaving them with no one fit to command—the replica of their situation when he came into it.

Of M. Jules Roy's much discussed brief book "La Guerre d'Algérie," published a fortnight ago, one literary critic here said, in grave appreciation, "It has taken six years of the Algerian war's death and tears for such a voice to be heard." Actually, two fraternal voices—one French, one Algerian—rise from its author, Roy being, like his friend Albert Camus (to whose memory he dedicates the book), a *pied noir,* or blackfoot, which is colonial slang for those white French born in North Africa. It is this dual identity, like a double slice of racial knowledge and human emotions, and this

French brain, familiar since childhood with the Arab mind, that give Roy's book its impact as the most moving condemnation to date of the war and the most authoritative simple explanation of its cause; together, these qualities have already brought it an enormous reading public at this moment of high tension in the Algerian problem. It was partly in memory of Camus that Roy felt forced to visit the Algeria of his youth this summer, where he crossed the damaged countryside "escorted by ruins and spectres"—the burned native villages and the roadside encampments of homeless, hungry Arabs. "I had to go to see it all in order to explain where it seemed to me the truth lay, provided I could find it," he writes.

His book's simple account, in the first person, is enriched by his feelings, mostly sadness and anger, at what he sees, is told, or remembers. A third-generation *pied noir* on his mother's side, he was brought up south of Algiers on her family's modest farm, where the men-folk had died, mostly of overwork or malaria, in clearing the marshland to plant their grapevines. They were helped to prosperity by the ill-paid Arabs, whom they called *les ratons,* the rats—a filthy, inferior, stupid, animal-like race (he was taught), whom, as he recently heard in Algeria from today's embittered, alarmed French whites, including his own sister-in-law, "our error has been to treat with too much humanity." Roy's first, and indeed final, conclusion about the truth, backed by pages of anecdotes and painful conversations with both races, is that simple "social injustice"—the white men's contempt, tyranny, psychological cruelty, and greed—"opened the road to the Algerians' rebellion."

The last part of the book could be called "Dialogue with a Captain," who is a French officer of harsh, high intelligence, a man of arms devoted to his cause with the passion of a crusader because he believes it noble and just—"the defense of the Occident," the great, civilized West. This dialogue is the crux of the book, a vivid, multiple-paged analysis by two military men of opposing political and social minds (for Roy served in the French Air Force for twenty-six years and fought as a colonel over Germany and in Indo-China). It is a basic dialogue between an Army *ultra,* or patriotic French Fascist, and a French intellectual liberal. It takes place dramatically in the captain's nocturnal fighting post, a control tower overlooking part of the electrified barbed-wire barricade that seals off Algeria at the Tunisian border so as to intercept and electrocute any Algerian night fighters slipping in from their camps in Tunis, though often enough it captures only wild boars or hares—"roasted, the captain said." In his finale, Roy declares, "The only way to stop this war is to negotiate—on condition that each of the adversaries abandon part of his pretensions." He ends his labors on the Algerian truths with a tender, reproachful invocation of his close friend Camus, who he thinks could have handled the matter better, saying, "You could have saved me all this by returning [to Paris from the country] by train on January 5th, as you wrote me you would, instead of leaving by car a day earlier." It will be recalled that Camus and Michel Gallimard, the son of his publisher, lost their lives in a motor accident on January 4th, en route to town.

November 15

Because of world tension, France's unrest, and the unclassic youthfulness of both the Democratic and the Republican candidates, our Presidential election last week aroused exceptional interest and worry among the French. The spectacular athletic-endurance quality of the last days of the campaign, the dramatic narrow margin of the victory, and, above all, the corporeal image that the contest set up in the active imagination of the French (themselves accustomed only to seated, elderly politicians), of two young men running a foot race for votes through the vast geography of the United States and ending, in a sweat, close to equal—all this stirred up almost as much spectator interest as political excitement among Parisians. What some of the French were worried about was money, of course—fear for their boom and prosperity, and fear of the effect that President-elect Kennedy's projected spending program could have on the dollar and the franc. In general, the well-to-do Parisians were the ones who spoke nostalgically of Vice-President Nixon; the modest-pursed ones seemed glad that Kennedy had won. Elderly Parisians—rich or poor, male or female—spoke as if they were convinced the United States had gone crazy in trusting its fate to a pair of such youthful politicos, with Kennedy, because he is four years the younger, regarded as exactly four years the less trustworthy of the two.

Until the Presidential election is over and won, Paris newspapers, as is their prudent quadrennial custom, print little that is personal about the two contestants except biographical material.

Then all the papers, each representing and influenced by its own special national and political inclination, produce their editorial analyses and portraits of the new incumbent. Of all the papers, last Thursday morning's *Figaro* gave the heartiest, most unexpected editorial welcome to Kennedy. As the leading morning guide of the *haute bourgeoisie,* it is ultra-conservative, old-fashioned, and pious—though supercritical, it is true, of the second Eisenhower regime. "Audacity and intelligence greatly aided John Fitzgerald Kennedy during the electoral campaign," it wrote, in part. "These are precisely the qualities that the thirty-fifth President of the United States most needs in facing up to the major problems that confront his country and all the West. The man is young, dynamic." Then it added, with an anti-Ike flourish, "The White House will find again the all-powerful quality it knew in the time of F. D. Roosevelt," and noted how easily Kennedy had "swept aside the old religious prejudices," without mentioning that the newspaper itself and the bases of those old religious prejudices are all Catholic. The liberals', the artists', and the intellectuals' modest morning paper, *Combat,* wrote, "Good luck to you, M. John Kennedy!" It then said, "The first ambition of this young chief of the American Democrats seems to be to cut loose from the geopolitical ignorance that in the past dictated to the Washington State Department its many awkward and unfortunate efforts"—a reference to the ghost of the unpopular Mr. Dulles. Then it criticized Kennedy, too, for thinking that America could help settle France's Algerian impasse. The extreme Rightist *Aurore,* fiery leader of the Algérie Française Army coterie and the colonial hotheads, ran its election editorial on Wednesday morning, addressed to "Monsieur le Président des Etats-Unis, whoever you may be, Nixon or Kennedy, Richard or John." It warned him that "your compatriots know little about Algerian affairs and are more susceptible than most to slogans, of which the latest is anti-colonialism," and advised him strongly to revise his opinions. What France was fighting for in Algeria, it said, was "a bastion against Communism," which could otherwise overrun the West "and you Americans, too." At the other political extreme, the Communist *Humanité* also addressed itself to both candidates, in an Election Day cartoon that showed each of them stuffing bundles of dollar bills into the ballot box. Its later editorial italicized its opinion that the United States had "voted not *for* Kennedy but *against* Nixon."

As for *Le Monde,* the dean of postwar French and international analytical political thinking, in the very first sentence of its Friday editorial it went straight to what it clearly considered the only real major point of the American election. It said, "The entire world is asking itself what consequences M. Kennedy's election may have on foreign policy." It stated that, at forty-three years of age, his ambition was not only to be elected this time but to be elected again in 1964. "The fragility of his Tuesday victory will thus force him to raise his own prestige, which the voters clearly let him know had not dazzled them."

The most personal French appreciation of Senator Kennedy's victory was addressed not to him but to his wife, née Jacqueline Bouvier, from the southern village of Pont-St.-Esprit, down in the Gard, where her French ancestors lived in the seventeenth century. It consisted of a long and surely costly cablegram sent to her by the village mayor and the municipal council, which offered "to the President and to yourself our warm felicitations and our ardent and respectful good wishes," adding, "The unanimous population of this little French town, the cradle of your family, salutes with enthusiasm and pride your entry into the White House."

The spectacular *procès des barricades,* or trial of the leaders of the January 24th Algiers insurrection, is now in its second week in the grandiose Cour d'Assises in the Palais de Justice by the Seine. The military tribunal on the bench is an imposing sight, with three red-robed civilian judges, and four generals and two colonels in medals and uniform. Their combined duty is to ascertain whether the sixteen accused men—four more are absent in flight—who crowd each afternoon into the prisoner's box, below a gigantic wall tapestry portraying some boy king of France on his throne, are guilty, as charged, of attempting to overthrow the French government and inciting to rebellion and involuntary homicide when they took up arms behind their Algiers barricades. Twenty-four people were killed, including fourteen gendarmes, and more than a hundred people were injured in their bloody riot against President de Gaulle's then new policy of self-determination for Algeria. Algeria is, of course, the complete essence of the trial—in the fact that these prisoners, in their way, represent all those other French who still violently believe or have believed that Algeria must remain a possession of France; in the fact that the courtroom houses the climax, at last and at law, of the dangerous division between de

Gaulle's progressing Algerian policies and those of the Algérie Française *ultras,* civil or military, and their repeatedly rumored coups d'état; and, finally, in the fact that, with three hundred witnesses to be called and a brigade of forty defense lawyers, all *ultras* (among them Maître Isorni, who was the defense lawyer for Marshal Pétain at his trial), the trial is bound to become a forum for anti-de Gaulle, pro-Algérie Française propaganda, and an attempt to expose the government to embarrassments, which have, indeed, already begun.

The trial opened with a Gallic bit of *opéra bouffe,* when one of the prisoners, against military orders, insisted on appearing in court in his insurrectional commando parachutist's uniform and was sent back to prison to change. Socially, the prisoners are a mixed lot of colonial Frenchmen, including a bank director, a *bistro* owner, several doctors, some insurance and real-estate agents, a professor (Comte Le Moyne de Sérigny, who was the editor of the *ultras'* newspaper, *L'Écho d'Alger*), and one lone Army officer. Wearing twenty-eight decorations on his pale-blue dolman, the intelligent, monk-faced Colonel Gardes, former chief of the Army's Algerian Fifth Bureau of Psychological Warfare, talked for a total of nearly thirteen hours during two afternoon court sessions, before crowds of lawyers who had nothing to do with the case but had come in from all over the Palais just to hear such a good talker. Among other strange items, he explained how he had been ordered by his Army chiefs to present de Gaulle's Algerian self-determination policy to pro-French Moslems and French *ultras* as "merely a maneuver to get around the United Nations debate on Algeria." The most important prisoner, the bearded, hawk-featured young lawyer-deputy Pierre Lagaillarde, regarded as the leader and hero of the barricades, where he stuck it out for eight days, was unexpectedly released on parole Wednesday, on a technicality concerning his Parliamentary immunity. Coming after two days of his brilliant, arrogant self-defense under cross-examination, the judges' sudden announcement of his release created pandemonium and one of the most melodramatic scenes ever witnessed in Palais history, with the defense lawyers and the *ultras* among the spectators screaming "Algérie Française!" in victory, and then singing the "Marseillaise" while the prisoners in the box stood patriotically at attention, after which armed guards rushed through the crowd, clearing the court. The government, even before Lagaillarde's release, which was considered a slap in the face for it, had already been badly embarrassed by the lawyer defending Dr. Bernard Lefèvre, who declared that the Doctor's part in the insurrection had been inspired by some May, 1958, writings of de Gaulle's Prime Minister Debré in the political review *Courrier de la Colère*. There he had written, "When the government violates the people's rights, insurrection becomes the most imperious duty —naturally meaning the Fourth Republic government, and naturally meaning the May 13th insurrection that brought de Gaulle to power. Under cross-examination, Lagaillarde had also greatly damaged the government's position on legitimacy by stating, "I took part in the attack against the government on May 13th, in uniform and armed"—naturally meaning the Fourth Republic government. "I am informing you that no punishment was ever prescribed. If you want to judge me for what happened on January 24, 1960, I demand that I be indicted for my activities on May 13th, which were of a much more highly insurrectional character."

On Wednesday, General de Gaulle also made his expected announcement of his proposal for a national referendum early in January on the new Algerian state, which he describes as an Algerian Republic, related to or independent of France, whichever the Algerians wish. In the incredibly aggravated present tension, many French now wonder whether planning anything in 1961 is not looking dangerously far into the future.

December 14

After the awful, bloody events in Algeria over last weekend, it would seem that the mere physical presence in that land of General de Gaulle as the symbol of France had, by its emotional power, tragically caused an explosion in each of the two races, the European settlers and the Arab natives—a freed combustion of their opposite reactions to France itself and of their hatred of each other. On Friday, the blood set flowing by the French *ultras* on the streets of Algiers and Oran, where de Gaulle did not even set foot, provoked no surprise, such rioting and killing of Moslems being one more of the *ultras'* customary dreadful messages to the chief of state—this time miles away across the sand in the little Arab village of Aïn Témouchent—that Algeria must remain French territory, in their masterful hands. It was the violent answer of the Arabs on Saturday, when they first swarmed out of their casbahs—the clamor of their voices by the thousands screaming "Independence! Ferhat Abbas to power! Free Algeria!" and the waving of their green-white-and-red Algerian rebel flags, these being actions and sounds never heard from them before—that furnished the terrified and total astonishment of what has been called in Paris "the day of truth." Until that day, through intimidation or understandable timidity (caught, as they have been, between the occupying French soldiers and their own *fellagha* army), the urban Arabs had never before demonstrated en masse what now seems to be their real national truth—their consuming desire for independence—for which their Saturday riot was their act of war. In Paris, it is considered that three myths died in Algeria over the weekend, these being the selfish myth of the white *ultras* that Algeria is French; the mendacious myth of the French Army that only a fistful of fighting rebels in Algeria wanted independence in all those years of war; and the major, miracle myth that de Gaulle could make peace—though no one here, or probably anywhere, thinks that anyone else could make it. He still stands alone, the unique dedicated figure of elderly courage and hope, as he appeared in Monday's photographs amid those pushing crowds in Algerian towns where assassination could have come in a gesture as easy as his handshake with those pressing around him—the burnoused Arabs trying merely to touch him, as if he were an amulet, while angry white French settlers brutally shouted across to him, "De Gaulle to the gallows!" Today, it is likely that only the Arabs still think that he can give them peace, as they also believe that he will give them independence—and, with it, new Algerian problems.

So far, only one criticism has been directed at de Gaulle, most respectfully made by *Figaro* in its Wednesday-morning editorial, entitled "The Return." He had come back to Paris the night before, and it was then impatiently hoped that he would at once speak to the nation on the radio, weary as he must have been—that he would say something, anything, because people wanted to hear his voice—and *Figaro* first criticized him for "not having uttered a word." Then it came to the heart of the matter, on which much of Paris public opinion agrees. "Too great an indifference to warnings and to the need for caution made de Gaulle's Algerian visit a tactical error," it said succinctly. "An enormous psychological drive had preceded Saturday's outbreak into action. A campaign of rumors and false alarms had been shrewdly spread by the partisans of one-way integration—that is, integration without equality or real fraternity between the Arab and the white communities. The overtones of it all announced that a maneuver was under way and predicted the coming of the tempest," timed to break upon his arrival. In conclusion, the editorial said, "In three days, the Europeans' rebellion against de Gaulle gave more aid to Ferhat Abbas than he had received in three years from the subversive rebel war." As you may recall, *Figaro* is the ultra-Catholic-bourgeois morning paper for the powerful, conservative middle class and the big industrialists, many of whom have heavy financial interests in North Africa. Until now, they have quietly backed the hope of a French Algeria, and this was the reason for their initial support of de Gaulle, in whom they then placed their faith. Such a *Figaro* editorial, including harsh criticism of the Algerian Europeans (whom it further called *"les comploteurs"*), has been regarded here as a strong patriotic and Christian warning to the *grande bourgeoisie,* and, coupled with the fact that certain Sahara oil shares fell heavily on the Monday stock exchange, has already caused a great deal of talk. Tuesday night's *Le Monde* printed—in a special box to attract the reader's attention, though without comment—the official report of the French government's Director of Information in Algiers, who stated that autopsies on fifty-six Arab corpses showed that of the twenty-nine killed by shots only five were killed by bullets of the sort used by the French Army or security forces, the others having been killed by pistol shots, which, according to multiple witnesses, including journalists, were fired by terrified French civilians on their balconies into the Arab mob below.

There is something cynically farcical in nearly all sudden French tragedies. This time, it was the government's Tuesday-night announcement that those responsible for the weekend riots and for the hundred and twenty-two dead (a hundred and fourteen of them Moslems, leaving only eight white Europeans) would be prosecuted. No moment could be less appropriate for such a project, since the continuing trial here of more than a dozen Frenchmen who took some part in the Algiers insurrectional riots of last January, in which a mere twenty-four people were killed, has so far resulted in a courtroom triumph for the accused. Pierre Lagaillarde, the barricade leader in those riots, and three other defendants fled to Spain upon their temporary release from the Santé Prison, as you no doubt know, and are at present safely out of the reach of French jurisdiction and justice. In fact, Lagaillarde's flight was openly considered here to be one of the disregarded warnings that accompanied de Gaulle on his recent journey to Algeria—although at least Lagaillarde was unable to arrive there for participation in the bloody weekend events.

Friday, the Parliament's current brief session—the second of the two sessions now permitted it in the year—will end. It is logical to suppose that the next few weeks will be decisive in Algerian affairs—and, indeed, in those of the Fifth Republic—though in what way nobody knows. Whatever occurs, the Parliamentary deputies can read about it in the newspapers, like the rest of the French, and with no more say-so or power. Tomorrow, Thursday, the Senate, which sits in the Palais de Luxembourg, was to have discussed general Algerian topics in a debate that had been scheduled before the riots. It is now thought that the government will decide that it is too soon after those tragic events to bring forth any official arguments—if only because of the lack, so far, of definite information for the senators, and of time for their mature pondering. Such a postponement would also enable the Debré government to avoid the imbalance caused by the absence from the discussion of the several Algerian senators, some of them Moslem, who have, with melancholy dignity, already refused to take part in any such talk. It is likely that the government will now merely read a statement to both the Parliament and the Senate—though here again no one yet knows what Premier Debré, recognized as an *ultra,* and one who writes his own speeches, may say. Too often before in this period of power, his sentiments have failed to resemble those declared by de Gaulle, and sometimes they have even been the diametrical opposite. Nor does anyone know whether the national referendum on Algeria that is scheduled for January 8th will actually be held then. Many citizens hope that it will not. It would, they think, be like reading the last will and testament too soon after the deceased has drawn his last breath. The Communist Party, which controls nearly a fourth of the French voters, announced a few days ago that it was conducting an extensive campaign among the working class generally—of any or all political affiliations—to bring in a massive vote of no against the de Gaulle referendum. In this, the Communists would be joined—or so it is thought—by the reactionary political parties that represent the Algérie Française *ultra* view. Some Paris commentators think that the *ultras'* position has been greatly strengthened by the Arab attack on Saturday, since it demonstrated the suddenly dangerous helplessness of the outnumbered French settlers there, whom France has always declared she would not abandon. The ultimate possible plan, which de Gaulle earlier proposed as a settlement of the Algerian problem—consisting of partition of Algeria, with the French, Italian, Spanish, Maltese, Jewish, and other foreign elements, lumped as Europeans, probably given the northern coastal ring of the country and the predominantly Europeanized cities, and the Arabs given some special territory, where they would be cared for and supplied with work and housing—is now considered, in the back of many French minds, as the worst that could happen. Separating the two races would be like separating capital and labor, Paris commentators think, or like cutting up the map of Algeria with a pair of scissors as a way of making history.

In the town of Blida, addressing a large contingent of young officers in training there, de Gaulle last weekend made his most informative, concise speech of the past half year, unfortunately little mentioned or quoted in the French press. Only the small intellectual morning paper *Combat* printed it here in full. In it he said, in part, "Because of this insurrection, the population of Algeria, which is Moslem by a great majority, has developed a consciousness that it did not have before. Nothing will prevent that. It must also be realized that this insurrection of the Arabs, with all that attaches to it, is taking place in a new world—in a world that in no way resembles the world I knew when I was young. There is, as you are well aware, a process of liberation now going on from one end of the world to the other that has affected our black Africa, that has affected all the former empires, without exception, and that cannot but have important consequences here. . . . It is thus necessary that the efforts of France in relation to Algeria continue, and it is very evident that they cannot be carried on in the conditions of yesterday. You can well imagine how a man of my age and my training has dreamed with regret of what could have been done earlier, and what was not done." It is to be supposed that the General, with his long historical memory, was thinking of the middle thirties in Algeria, when the only rebellion imagined by the Arabs was to ask for integration on what would now seem modest, loyal, republican terms. They asked to be integrated into the French nation with the privileges of the vote and representation accorded to all French citizens—the privileges enjoyed by the white settlers there, who refused them.

It is said that President de Gaulle will probably address the nation tomorrow night, though this has already been denied by a semi-official rumor that he will not speak until next Tuesday, or even sometime in Christmas Week. The nation is waiting—with patience, if necessary—to hear that unique, broken, trumpeting voice, and to listen to his thoughts.

December 28

For many years now, monthly bulletins on economic and social problems have been sent out to the press here (which for various reasons has rarely used them) by the Bourbon pretender to the throne of France—Henri, Comte de Paris—who is reputed to be Socialistic and is known to be pro-de Gaulle. His bulletin of December 14th has exceptional news value, because in it he announces that he is suspending these studious, rather solemn communications, in what he calls the troubled times France is going through, and says why. It is, he states, because "I trust General de Gaulle." With close to royal dignity, he declares that "rather than uselessly mingle my voice in the frantic clamor, I, whose vocation is to unite and pacify, choose to keep silent while waiting for the day when the French will comprehend" that they must unite "in understanding of and respect for the realities of our time and the world we are living in." To the *colons* of French blood in Algeria, he sternly adds, "Follow General de Gaulle. He is the only chance France has, your only safeguard." With this he stirred up a hornets' nest among what is left today of the old ultra-reactionary, anti-republican, onetime pro-Pétain, Action Française royalists. In their weekly paper, *Aspects de la France,* they declared with anger, "We have a right to be consternated by the Prince's terms, which constitute unconditional approbation of General de Gaulle" and a deplorable acquiescence in "the error of modern times" by the "inheritor of the Capets" (his earliest dynastic ancestors, who rose to power around the year 987). After describing Algeria as "the last land tied to the crown by our kings" (meaning the unlucky Louis Philippe), they end up by saying frankly that "it is as much for the heritage of the present pretender himself as for France's immediate interests" that the Action Française "refuses to follow M. de Gaulle into Algerian Algeria." All these anachronistic references to kings, addressed to those monarchs' final, symbolic, Socialist relic, the Comte de Paris, and coming to him, as they do, after the machine guns and Molotov cocktails on the streets of Algiers a fortnight ago, make it clear that President de Gaulle's problem of the mixed races in Algeria is also a problem of the persistent admixture of certain kinds of strange French people living in France today.

1961

January 12

It is easy now, after the success of the event, to say that President de Gaulle was bound to win his majority in metropolitan France in the referendum last weekend, though actually those were three of the most confusing days, involving conscience, dialectics, and muddled opinions and hopes, that French voters have gone through in all the cloudy political years since the peace in 1945 began. It is easy to say that he won by staking his reputation on the outcome and by even raising his ante with a faint threat to resign if he was not given the "frank and massive 'yes'" that he said he must have in order to carry on his Fifth Republic, with no one in France suited to fill his shoes were he to abdicate. Yet there was certainly an incalculable but important number of bourgeois citizens who, though they forced themselves to vote yes, are politically tired of the abnormalities of personal power essential to his regime and who openly look forward to the end of the war in Algeria as also the end of de Gaulle, that extraordinary leader of an interlude, after which France can get down to the necessary, more vulgar reality of trying to run itself again. It is also easy to say that he won his unexpected 69.09-per-cent affirmation in Algeria because the French Army of Occupation was ordered to tell the Moslems to vote yes. Yet the French Army itself largely voted no. Just before the referendum, sixteen generals signed a public round robin of protest against their chief's Algerian Republic policy, declaring their devotion to the French Algeria that they had all fought for and still believe in. France's only living marshal, Maréchal Alphonse Juin, who was born in Algeria, even wrote a rebellious, insolent letter to General de

Gaulle and gave it to the French newspapers to print. De Gaulle's sole answer was to remove Maréchal Juin from the Supreme Defense Council and forbid him to go home to Algeria to propagandize for a vote of no. He also sacked General Salan from his high-council functions, so Salan is now sulking idly in Spain. The more important of two secondary de Gaulle victories in the referendum is its proof that though he may well have broken the hearts of the Army's élite, he has also broken the Army's political power. His third referendum triumph lies in the ludicrous collapse at the polls of the previously self-important Right Wing Algérie Française diehards here in France. They voted no, in a political paradox, with the Communists, the other brash talkers, yet even together the two parties polled a no vote of only 24.74 per cent. Of all the political parties, the Communists had laid down the heaviest barrage of propaganda; in the Paris working-class districts daily bundles of tracts from *L'Humanité* in favor of no were stuffed into voters' mailboxes.

De Gaulle's final television appeal to the voters to let him "know what is in your minds and hearts" had some unexpected and funny reactions. In the wine regions of the southwest, certain groups let him know they had their minds made up against his recent viticultural policy. In Corsica, where there was talk of the Paris government's suppressing a small railroad, the peasants let him know that in their hearts they were certainly against his stand on the railroad question.

As a result of the referendum, de Gaulle's policy in Algeria—his plans to create a provisional Algerian government of Moslem Algerians even before peace is obtained or the Moslems' vote on self-determination has been held—is now law. In this week's meeting with his Ministers, he has already approved the creation of Algerian administrative and deliberative institutions, so that the Moslems will have had some practice in doing things for themselves when the great day of freedom finally comes. What the referendum did not mention, but what surely lies behind all the yes votes, is everybody's desire for a negotiated peace. The sooner General de Gaulle arranges for it, the broader his historical fame will lie, on both sides of the Mediterranean.

Edith Piaf's triumphant and pathetic recent opening night at the Olympia Music Hall was regarded as a resurrection. For two years she has been in a kind of limbo of mortal bad luck and danger, beset by illness and motor accidents, her money prodigally wasted, her apartment stripped to pay her debts, all this followed by a collapse on a provincial stage when, of necessity, she started to sing too soon, and, finally, the fear that her voice was gone, like everything else. Since she is the best loved and most significant of the *populo Paris chanteuses* (her mother reportedly gave birth to her on a sidewalk in Belleville), she and her melodramas were followed with that special sympathy the Parisian public gives to its aging theatre favorites. Her first night at the Olympia was a gala benefit performance for the First Free French Division, which helped liberate France. Old General Georges Catroux was present, as well as some Cabinet members, various stage and movie stars, and the Garde Républicaine, in their helmets and white breeches. Her triumph has continued nightly, with the crowd shouting its appreciation and battering her fragility with its unrestrained applause. What makes her performance touching—and, in a way, terrifying—is that her tiny, stiff-moving body and her lunar, modest face, with its faint nimbus of sparse hair, belong to an unfamiliar, ravaged woman inhabited by that familiar voice, which suddenly bursts forth without age or change and with its rolling, throaty "r"s, singing about disappointed love like a city-park nightingale. She walks with difficulty, her heavy feet shapeless in sandals, and her poor-looking black dress (her trademark since her young days, when she sang and begged on the streets) stretches tight over her thin arms and chest. On legs spread wide to give her balance, she fixes herself before the microphone, and then comes her soaring, intact sound, carrying her ballads—every banal word and the full drama of their meaning perfectly articulated—without a single physical gesture, not even that of lifting her hands. When the thunderous applause strikes her, she mostly acts as if she did not hear it, or else vaguely smiles, being in haste to announce the next song and get on with her heavy program. It consists of ten new songs and, as final encores, seven or eight of the old favorites—a champion performance.

Her contract at the Olympia is for a month. People queue up all day, every day, on the boulevard in front of the ticket office. Nightly, she brings down the house with her confessional new song about her own life, "Je ne Regrette Rien," with its plucky refrain, *"Balayés les amours, avec leurs trémolos, Balayés pour toujours, je repars à zéro!"* In one song, called simply "Mon Dieu," she asks God to give her a few more days of happiness with her false love. In "Boulevard du

Crime," a Guignolesque ballad, she essays a few mocking dance steps. "Les Blouses Blanches" is about the white-jacketed nurses in a madhouse. Her new sentimental successes are "Les Mots d'Amour" and "L'Homme Qu'Il Me Faut"—typical nostalgic Paris *bistro* waltzes. Artistically, her dramatic projection of her songs is still unequalled. Starting again at zero, she is making a new fortune for her new self. Her real name has long been forgotten—Edith Giovanna Gassion. Since her teens, when she first began chirping her songs, she has been known as Piaf, which is Paris slang for "sparrow."

January 24

President Kennedy's Inaugural Address stirred rare interest and praise here. It was talked about with satisfaction by Paris citizens and was written about with serious appreciation by the Paris papers (one of them called it "an exception in the Western political literature of today"), and the laudatory reaction to it was the same whether the people or the newspapers were right, left, or merely middle in their own local politics—a unanimous frame of mind extraordinary in France. *Figaro* devoted a special editorial, entitled "A New Tone and Style," to what the President said and how he said it, pointing out that "his formulas were constantly direct and his figures of speech always striking, making no mystery of the dynamism, which youth alone cannot explain. . . . It was virile language, aimed at arousing the energies of a great nation that is menaced today by the very excesses of its own prosperity." *Le Monde's* Washington correspondent, a Frenchman, cabled that "all the American and foreign confreres agreed in thinking that his speech was of an exceptional richness of thought for a discourse of this sort," adding, "It was written in beautiful language, in which certain phrases resounded like poetry."

February 9

All France is once more waiting for the decencies of peace. The French had hoped that de Gaulle's referendum majority last month would lead to a negotiated peace as directly as might be possible for the stately de Gaulle and the Oriental-minded Moslems, neither side being likely to proceed by nature in a straight line toward any important settlement. Now we hear that each side says the other has got to start first on the humble pie of opening the negotiations. On Monday, everybody was hailing the Paris headline news, taken from a Tunisian weekly, *Afrique-Action,* that President de Gaulle, of France, would like to see President Bourguiba, of Tunisia, in Paris, at the latter's convenience—the eager interpretation being that Bourguiba, who needs some extra French good will on the question of the port of Bizerte, might serve as the honest broker between Algiers and Paris in peace talks. He has been quoted by *Afrique-Action* as saying that General de Gaulle is understandably feeling his way, is looking around, is taking precautions that may seem superfluous. "But it would be to nobody's interest to have him run too many risks or have him fail," Bourguiba is reported to have said. "That would be a catastrophe. And if three or four weeks from now the peace negotiation has not got under way, that will be an equal catastrophe." On Wednesday, M. Mohamed Masmoudi, Tunisian Minister of Information from Rabat, was received at the Elysée Palace by President de Gaulle in "a cordial conversation," though what about was not said. It is deduced that M. Masmoudi will next get in touch with President Bourguiba, now on holiday in Switzerland, who will get in touch with Ferhat Abbas, now in Cairo, who, as president of the provisional Algerian government, is the rebel in exile with whom the peace must be eventually made. Then M. Masmoudi will arrange for Bourguiba to have "a cordial conversation" in Paris with de Gaulle on how the land lies. But nothing so far has been said to the French public on any of these vital matters by General de Gaulle. It is a curious experience to be living under what is so often and for so long a silent government, led by a man who is the greatest speaker now in power in our hemisphere.

February 24

Napoleon III, though he has never been considered a very farseeing monarch, had an early, precise metaphor for Algeria, which around 1863 he called *"un boulet attaché aux pieds de la France"*—a ball and chain attached to France's feet. "Algeria is not a colony, properly speaking, but an Arab kingdom," he added, in a testy spirit. With the French government's sudden announcement yesterday that President Bourguiba of Tunisia, expected here last week, will positively appear on Monday to talk with President de Gaulle about de Gaulle's coming talks with Ferhat Abbas, President of the Algerian Provisional Government, on peace in Algeria, it looks as if the colonial ball and chain were going to be soon removed. The protocol of this Monday visit is worrisome, since Arabs are touchy. However, the visitor's safety from death has to be taken into account as part of the courtesy he is entitled to. Personal violence against him from embittered French *ultras* here is naturally to be feared. Where to house him safely and how to treat him socially so that he will not have a bomb thrown at him are matters that have apparently not yet been settled. He is a visiting chief of state, so he must be received as such, but he must not be entirely treated as such, with a parade and parties, which could be mortal. The French government had thought to house him in the magnificent Château de Champs, only twenty kilometres from Paris, which was built in early 1700, was occupied eventually by Mme. de Pompadour in tremendous style, and during the nineteen-thirties, when it was owned privately, was the scene of some of the grandest balls of that giddy period, after which the state got it back as a gift. But already the Tunisians have politely let it be known they prefer not to be quartered in country quiet but wish to be handier to things in Paris. The memory of how the Algerian peace delegation at last summer's fiasco meeting in Melun was cut off, as if in exile, from all social intercourse, and even from telephone communication with Paris, still rankles in all Arabs. In addition, the North African Arabs have not only the embittering long recollection of colonialism but right now the political hypersensitivity of the dark-skinned, brought into focus by the various kinds of explosions between white and colored all over the globe, from New Orleans to the Congo. The most difficult diplomatic, policing, and housing problem will be that of where to put Ferhat Abbas when he comes, for he is known in advance to be a morbidly sensitive, patriotic Arab and a high symbol of the French *ultras'* hatred. And will President de Gaulle, with his genius for French glory and Gallic formality, actually consent, after all, to "have direct contact with the chief of the Algerian rebellion"—even to make peace? Once again this question is being raised here with fear. Along with Thursday's brightening news about Bourguiba came the bad news from Algeria that suddenly the rebels had picked up the war again and were fighting, at exactly the wrong historical moment. Yet the war, of course, is really over, except for occasional terrorist murders or bombings here and there—and except for talking about, making, and signing the peace.

The largest, most illuminating collection of Henri Rousseau's paintings ever shown in Europe is now on view at the Galerie Charpentier. Consisting of eighty canvases, it is the first major Douanier exhibition in Paris since the two retrospectives given him shortly after his death, from gangrene, in the Hôpital Necker, at the age of sixty-six—the one at the Salon des Indépendants in 1911, and the big commercial retrospective put together in 1912 by the art merchants Bernheim-Jeune, who had collected a valuable lot of Rousseaus in the two years after the old man was buried in a pauper's grave. Among the notable items at the Charpentier are nine jungle scenes, three of which, owned by Paris collectors, the Paris public had apparently never laid eyes on before, any more than it had on the privately owned gray-colored lion's face and the two full-length owls (these three look like miniature animal portraits), making six brand-new aesthetic pleasures in all. But even more talked about is the new, serious, appreciative attitude that is being manifested by Paris art critics and art lovers toward the Douanier as an artist, and the completely new inquiry that is being made into his character. Formerly, he was viewed as a lovable simpleton who was arrested two or three times for minor thefts, probably a mythomaniac but possibly an actual ex-soldier who had accompanied an expedition to the jungles of Mexico, and the admired innocent painter of the popular wedding-party picture "La Noce" and of the monstrous barelegged child with her doll. In last week's *Arts,* which devoted three full-size newspaper pages compiled by four Rousseau experts, and accompanied by fifteen photographs to reveal the new justice being done him and to present the now available truths about him, a statement by that weekly's M. Yann Le Pichon was the most astonishing (including its failure to explain why the revelation has come so late). Le Pichon says that just after Rousseau died, his daughter, Mme. Julia Bernard-Rousseau (the only one of his nine offspring to live to maturity), found in her father's Rue Perrel studio a battered children's book on animals, which she took for her own child Jeannette—a cheap, dog-eared volume she later showed Le Pichon. The cover of this book, a publication of the well-known department store Aux Galeries Lafayette, was reproduced in *Arts,* and showed an appealing design of lizards, snakes, and tropical birds surrounding the title: "Bêtes Sauvages: Environ 200 Illustrations Amusantes de la Vie des Animaux, avec Texte Instructif." Inside were good realistic drawings of monkeys, whose exact poses on the trees Rousseau zealously copied in such paintings as "Jongle" and "Paysage des Tropiques," though he managed to have it appear that the monkeys were fancifully playing ball, like little boys, with oranges. The album's serpents and duck-billed birds served as models for his magical "Charmeuse de Serpents," also present in the Charpentier exhibition. The department-store album's lions and tigers Rousseau set loose in his imagination amid the Paris Jardin des Plantes's exotic botanical exhibits, whose salad-green or elephant-black foliage he tenderly, laboriously copied, leaf by leaf, for he was an uneducated classicist. Le Pichon's disclosure about the animal book, together with certain references from an Army report on Rousseau, now proves beyond a doubt that there never was any adventurous journey to Mexico with Maréchal Bazaine's troops, which Rousseau fondly pretended was the source of his jungle paintings. He voyaged only on his pictures, as on magic carpets.

In the new evaluation, which changes Rousseau's aesthetic ranking, he is hailed as having created his own clear, logical painting line in a reaction against the spongy softness of Impressionism. He is credited with being the first important French exponent of *l'art brut,* now much admired here for its oneiric, providential creativeness, and he is also called the forefather of Surrealism. As for his most naïve, puerile canvases, which the public has so affectionately appreciated, they are now said to contain, aside from their coloration, "nothing admirable." Quite rightly, the revelation of the Charpentier exhibition has been his hitherto little-known Paris water-and-land paintings, as examples of calm, minor French painting genius, such as the 1888 "Evening View of the Ile Saint-Louis and Bridge," the 1909 suburban "Fisherman Among Trees," and the 1893 "Sawmill Outside Paris," with its humble but superb composition of pallid logs on a curved forest road, along which a mother walks with one child in her arms and another at her side—a great and human small painting.

In the matter of the mind of the Douanier, a man until recently regarded as the helpless butt of everybody's practical jokes—proof of both his naïve nature and his naïve art—the new criticism has a tougher problem of reappraisal. The painter Gauguin and the writer Alfred Jarry, author of that unrefined satiric play "Ubu Roi," were his liveliest, most inventive tormentors. They privately presented him with the Legion of Honor, which he blissfully wore; hired a tramp to dress up as Puvis de Chavannes and call on him with words of praise; sent to visit him the real painter Degas, of whom he mildly inquired, "And how is your painting coming along?" Rousseau told Cézanne, apropos of the latter's Salon des Indépendants paintings, "I could finish those for you"—for, after all, Rousseau had been showing at the Salon for years before Cézanne was accepted. One night, Gauguin told Rousseau that he, the Douanier, was expected by the President of the Republic at the Elysée Palace, and Rousseau washed and hurried over. When he returned, he said that the President had received him most kindly but had explained that everyone else was in evening clothes, so would he please come back another night. True? Or was he, in turn, pulling the malicious Gauguin's leg? The shrewd, hard Vollard, his art dealer, at times suspected that Rousseau was really as sharp as a tack. Many people regarded him as an enigma, but some thought that he imaginatively played the fool to fool others—his complacent form of fun. Yet he referred to all foreign painters as Americans, which was surely irrational, and he shouted over the telephone because the people he was talking to were "so far away." It is now supposed that the famous Montmartre banquet of honor given him by Picasso and Apollinaire at the Bateau-Lavoir was only an elaborate, vinous farce. Since they so wonderfully built up his painting reputation as a joke, it is further deduced that once he was dead and they had the legend of his innocent genius laid at their door, like an exotic funeral wreath, they had to revivify it, out of shame and loyalty. Three years after his death, his painting "The Virgin Forest," which he himself had vainly tried to sell for two hundred francs (then forty dollars), fetched ten thousand gold francs. The legend had become truth.

March 8

It is to be noted that much too little has been said overtly by anybody about the Sahara oil. Ten days ago, Premier Debré, on a Sahara tour, said, as if to the desert air, "Take note that France is present here and that she will stay." At once, the Algerian rebel government reaffirmed its earlier stand that the Sahara "forms an integral part of Algeria" and that "sovereignty must be exercised over the desert as over the rest of the national territory." To that western part of North Africa, the Sahara oil is almost what the Katanga mines have been to the Congo.

April 5

No one now knows where the peace negotiations between the French and the Algerians are headed for, or have gone to. Surely never in modern European history has the comic fiddle-faddle of squabbling officials, supposedly with olive branches in their briefcases, seemed more melancholy, more sad. Plastic bombs believed to have been planted by white Algérie Française *ultras,* who are against the peace ever coming at all, killed six people in France and Algeria over the Easter weekend and wounded fifty, with thirty more wounded this week by a plastic bomb in the men's room of the Paris Bourse. The Hôtel Lutétia received a letter of menace because M. Mendès-France was scheduled to speak there on Wednesday afternoon to the press, but nothing untoward occurred, nor did he say as much as was hoped, except that "the little diplomatic guerrilla war must cease, since it is making the war itself last longer."

April 27

What we have had here has been the alarms, fears, and repercussions of a longish-weekend Algerian military insurrection, fomented on the other side of the Mediterranean by four retired French generals, which began as a total surprise to the French government shortly past midnight on last Friday night and ended as a completed failure, amid the surprise and intense relief of the French people, a little past midnight on Tuesday night. The whole restless, ill-focussed adventure had a crazed, nightmarish, nocturnal quality for all of us here—especially, of course, for the French, because all of the few events that connected them with it took place at night, with one major, miraculous exception, which took place on Monday afternoon. In it, the entire French nation woke up to a spirit of rare, refreshed unanimity against the folly of what was going on—millions of French all over France, clear-eyed for once, as if coming out of an insanely grotesque dream, joined in the greatest general strike of protest of all classes, political beliefs, and social levels that France has ever known in all its agitated modern history. It simply consisted of everybody's stopping work early, at five on Monday afternoon, thus forming what seemed like a vast human blockade in the path of the far-off Army plot—a block of French republicanism loyal to its government and its leader, whether or not it ordinarily likes it or him, in a sudden period of danger and stress. Here in Paris, the Métro, the buses, and the commuter trains stopped running, movies stopped showing their films, shops closed, the post-office clerks laid down their stamps, everything came to a standstill. It was a handsome sunny spring afternoon, and everybody, armed with the latest newspapers, took to the midtown boulevards to stroll, almost with insouciance. Only the cafés remained open for work, but too crammed to do business. The Place de l'Opéra was a mass of human beings, gay in the open air, as if having a fête. It is now known that this gigantic inhospitable demonstration against the idea of parachutists' dropping in from Algeria to dismantle France's Fifth Republic by a military coup added to the confusion of the four seditious generals in Algiers, whose *Putsch,* whether they had sense enough to realize it or not, had already failed. It was these four traitors who were leading this new *Putsch.*

The opening evening event had occured Friday, after President de Gaulle had been characteristically attending a performance of Racine's "Britannicus" at the Comédie-Française with a suite of his Ministers, including M. Louis Joxe, his Minister for Algeria—thus proving that the French President's Secret Service for Algerian affairs was as far off the track on insurrection matters as President Kennedy's C.I.A. had been on Cuba. At two o'clock in the morning of Saturday, de Gaulle was awakened in his palace to be given vitally important and alarming information: four of his former comrades and admirers were now leading the third Algerian insurrection to force the hand of Paris. The first insurrection on May 13, 1958, had pulled down the Fourth Republic and brought de Gaulle himself to power, and at that time General Raoul Salan had actually raised the first voice to call for de Gaulle, who later named him Commander-in-Chief of Algeria. General Maurice Challe, a notable fighter and organizer in de Gaulle's wartime Resistance, was later also named Algeria's Commander-in-Chief, but was replaced after the Algerian-barricades insurrection in January, 1960. Algerian-born General Edmond Jouhaud had also been a great Resistance fighter, and was made de Gaulle's Chief of Staff of the Air Forces in Algeria, but later declared that he would vote no in the January, 1961, referendum on de Gaulle's Algerian policy. The fourth and least consequential, General André Zeller, an up-from-the-ranks volunteer from the First World War, was chiefly distinguished for being the most loquacious rebel of all the Algerian Army brass against de Gaulle's policy.

Sunday night at eight, General de Gaulle finally addressed his nation, on television and in uniform. It was the greatest speaking performance of his career, being the words and voice of an aged patriot, ripe in civilization, wounded in heart and mind, angered by the treachery of former friends, emptying the classic phials of his disdain upon the evil "usurpers—partisan, ambitious, and fanatical—who see and comprehend the nation and the world only through the distortion of their frenzy," in an outburst of scorn as old as the antiquity of power itself. When he cried three times *"Hélas! Hélas! Hélas!"* it was the male voice of French tragedy, more moving, because anguished by reality, than any stage voice in "Britannicus." "The state is flouted, the nation defied," he said. "In the name of France, I order that all means—I say all means—be employed on all sides to bar the route to these men until they be subjugated. I forbid all Frenchmen, and first of all any French soldier, to execute any orders of theirs. . . . *Françaises, Français,* look where France risks falling, compared to what she was once more about to become! *Françaises, Français, aidez-moi!"* Then came the "Marseillaise," trumpeting at the call *"Aux armes, citoyens!"*

Just before midnight, when the state radio is always turned off, a voice announced that it would function all night, because there might be grave news to impart. A little later came the bourgeois voice of Premier Debré, speaking jerkily and in detached phrases, to make them plain, and declaring that "a surprise action, particularly in the Paris region," was shortly expected—"a mad attempt" by aircraft "that are ready to drop or land paratroopers on various airdromes to prepare for a seizure of power"—and that, as of midnight, all French fields were closed to airplane traffic. Then he added, as one of the queerest of all the nocturnal experiences, "As soon as the [air-raid] sirens sound, go there on foot or by car to convince the misled soldiers of their grave error. Good sense must spring from the people's soul, and each must feel himself a part of the nation." This speech was still alternating on the radio with de Gaulle's at 3 A.M., when many of us millions of worried, worn-out listeners went to bed. At least one of us (your correspondent) was possessed by an imaginary *tableau vivant* in which we saw crowds of hastily dressed French citizens on foot at some airport convincing misled, tough Foreign Legion paratroopers—eighty per cent of whom are Germans—of their erroneous conduct.

Next morning, Monday, there was a cynical, superior tendency on the part of a few—which soon became the day's vogue—to say that Debré's speech was a bluff, that there had been no airplanes ready to take off, that the bluff was a cracking good piece of psychology that someone must have thought up for him in order to arouse and unify with terror the usually apathetic French. At the present writing, no one knows whether this is true or not. All over town, people say, but no officials have admitted, that the insurrection's planes were indeed ready to fly up toward France on Sunday night but that some pilots refused to fly against the homeland and some French parachutists refused to go aboard. Other Parisians say that there was a bad Mediterranean storm that held up flying; that some planes were sabotaged; that they had a flying radius of only a thousand kilometres, or not enough to make Paris and return. And so on. What is definitely known is that France was naked of protection on that Sunday night, when four small tanks were clustered as defense before the President's Elysée Palace, plus policemen and Gardes Républicaines on foot—a chief of state in complete vulnerability had his palace been invaded. Thereafter, big tanks were nightly shuffled into position before the Parliament, with a soldier or two, their heads in the green leaves of the Quai d'Orsay trees, asleep on top in the steady night rains. Empty prewar autobuses were lined up nightly in the Rue de la Paix, on the bridges, in the side streets near de Gaulle's palace, to serve as impediments in case the paratroopers came to town. And in the rain, all of us in Paris went to stare at them. Each morning, they were all cleared away, both tanks and buses, at breakfast time, in a gesture like emptying dustbins that during the night have accumulated unsightly social debris. All this time, everybody knew that France had half a million men, including the Foreign Legion mercenaries, under arms, more or less, in

Algeria, but that she had no men worth speaking of here on the home ground, and could not be sure even of their loyalty or that of the special shock police. The insurrectionists had a wealth of equipment in their favor, but they lost the chance of their greatest imponderable—that of surprise—when they failed to descend from the rainy skies on Paris Saturday night.

Throughout the painful long weekend, before the insurrection suddenly collapsed and the generals fled in ignominy, the question constantly asked, in various phrasings, was "How can educated, more than middle-aged, highly trained Army men have launched themselves on such a crazed adventure today, in the face of the world's risen tide of liberties for all?" One possible answer is that, forgetful of loyalty to de Gaulle and duty (for which derelictions all four generals had been earlier reprimanded or removed from their Algerian high posts), they had imbued themselves with the white Frenchman's sense of possession of Algeria, like those old-fashioned figures in Delacroix canvases, galloping in proud ownership across its exotic scene and sands.

May 2

Last Friday afternoon, five days after the insurrection plot against President de Gaulle and his Fifth Republic by the four ringleader French ex-generals, *France-Soir,* which is the biggest popular paper in all France, presented a front page that looked like a bankruptcy report on the morality of the French Army and the French civil and public services. The facts and figures, printed in round numbers and adding up to hundreds of Frenchmen in high or trusted or secret places who had been accused of complicity in the mutiny, were more frightening to most of the French public than anything in the four days of the rebellion crisis itself. For nothing really happened here during those four days, so that fear became vague and empty of reality and was replaced by confusion, deep worry, and a kind of angry curiosity as to what was really going on. *France-Soir's* front page was something positive and informing, which the eye could take in at a glance—something enormous, only a fraction of which we had supposed or suspected. It was a printed opening total, to date, of the top Frenchmen, often listing their names as well as their posts, who had been involved in what still seems to have been an insane *Putsch* to overthrow de Gaulle's government, take power, and rule the next Republic of France by military junta. The front-page announcements were mere headlines, big and little. They stated that General Marie-Michel Gouraud, former commander of the Constantine Army Corps, had been imprisoned, along with four other generals and five colonels; that two hundred officers in the French Army had been arrested; that four French Army regiments had been dissolved, including the famous, overpopular 1st Regiment of Foreign Legion parachutists, whose acting commander, Major Elie de Saint-Marc, had also been arrested; that the 14th and 18th Foreign Legion Regiments and a commando regiment of French paratroopers had likewise been dissolved. They said that the second most important officer in Algeria—General Héritier, chief of staff of the combined Army, Navy, and Air Forces in Algeria—had been removed from his post, and also Colonel Cousteau, chief of the Troisième Bureau, which plans military operations. In Paris, three officers from the Ministère des Anciens Combattants had been arrested, the paper said, and so had a hundred functionaries and four hundred members of L'Organization de l'Armée Secrète in Algeria—where the chief magistrate of Algiers, the Président de la Cour d'Appel, had himself been suspended. General Jacques Faure and the ten officers who were among the first arrested down there have now been questioned here by inspectors of the Criminal Brigade of the Quai des Orfèvres—an office and a police force made famous by Georges Simenon's Detective-Inspector Maigret, who, in the ordinary way, would never have met eleven high Army officers unless they had all become murderers or bank robbers. It is to be noted, too, that ex-General Maurice Challe, official leader of the ex-generals' quartet, who gave himself up and will shortly stand trial for his life, has been lodged in the common-criminal division of La Santé prison, not in the more select cells for political rebels.

On the other hand, amid all this appalling news, there is word that General Guy Grout de Beaufort, who held the critically important high post of Attaché de l'Institut des Hautes Etudes de la Défense Nationale, here in Paris, and who had earlier been chief of de Gaulle's own staff, has been released from arrest and is now only under surveillance. However, General André Petit, formerly Premier Debré's military adviser, has just been imprisoned, and his successor, General Jean Nicot, of the Army of the Air, who was the officer responsible to Debré for France's air defense, has just been put under fortress arrest. He is accused of having enabled, possibly with forged passports, two of the plotting generals—Challe and Zeller—to secretly leave France (where they had been restricted by government order, because of their stated disagreement with de Gaulle's Algerian-independence policy) and to regain Algeria and there lead their crazed, inefficient mutiny. In all, fifty more arrests were made in Paris over the weekend, bringing the total to about three hundred and fifty for all France since these stern repressions began—a few weeks and several years too late. In damage to France's morale and international standing, this has been the most disastrous plot of Army disobedience to state authority and the most enfeebling split in French Army unity—in theory still a sacred *mystique* here—since the notorious Dreyfus case.

There is embarrassment and real grief among many Parisians over the disloyalty and dissolution of the famous 1st Regiment of the French Foreign Legion. In military parades on the Champs-Elysées in the old days, before they were paratroopers and took to the sky—when they still carried their earthy pickaxes or shovels tilted over their shoulders, like weapons for their peculiar combat with the desert sand—they always received the greatest storm of applause from the Paris sidewalk crowds of any unit except France's own élite sons, the cadets from Saint-Cyr. The Legionnaires' snow-white kepis, their long, slow stride for desert walking, the fringe of beard that most of the *sous-off'* wore on their stony bronze faces, their unified air of perfected masculine combative discipline, like homeless fighting cocks, endowed them with a spectacular attraction that magnetized the domesticated French bourgeoisie. The Legionnaires' melodramatic departure last week, when they blew up their quarters at Zéralda, so no other soldiery could be sheltered behind what had been their own walls; the civilian crowd from Algiers, for whom they were the supreme favorites, tossing red roses and bottles of cognac as forms of farewell into their trucks; and the Legionnaires pulling out while singing insolently, *"Je ne regrette rien, je repars à zéro,"* the Piaf hit *chanson* this winter at the Olympia Music Hall—all this was part of their bold and inspiring theatricalism.

Nearly nothing has been told here in Paris about how the insurrection got under way in Algiers on that Friday, April 21st. Information has just been gathered by a couple of frequent visitors to

Algiers, now returned, who flew down there from Paris a few days after the insurrection's collapse to listen to their many Algiers acquaintances and friends—Army, civilian, political—and also to French settlers, both pro- and anti-Algérie Française. Some of the facts they returned with seemed absolutely incredible and were merely perfectly true. According to these informants, events started on Friday afternoon when a soldier in the Kabylia district told his captain, "I have been ordered to go to Algiers tonight to take part in a *Putsch*." The captain told his major, who told his colonel, who told his chief, General Simon, who hurried off in a helicopter to tell the Army corps commander, General Vézinet, in Algiers, where the news, still going up the ladder of command, was passed to the top echelons of the Air Force and the Navy, and then to General Fernand Gambiez, the Commander-in-Chief in Algeria of the combined Army, Navy, and Air Forces. The first reaction on a high level was that this was merely another case of *l'intoxication* (usually called simply *l'intox*), the maladive hysteria of rumors common in the always tense situation in Algiers and caused by the close quarters of the different races and the military forces. At eight that night, a meeting was held in the office of Jean Morin, Delegate-General of Algeria, attended by all the brass—General Gambiez, the directors of the military and civilian cabinet, the head of security forces, the prefect of police. Somebody produced *une lettre confidentielle*—one of those extreme-right-wing newsletters (its author has now been arrested) that can be subscribed to for a fee in Paris—which said, "At 2 A.M. Saturday morning, the loyalty of the Army can no longer be counted on." This sounded to the brass like sheer bosh; it was unbelievable that any insurgents would let out semi-public news of their D Day and H Hour in advance. But patrols were ordered doubled and the 14th Squadron of the Gendarmerie was alerted. Then the brass played some bridge and went home to bed. Before midnight, somebody reported by phone to the general staff that a transport truck company had been seen en route to Zéralda. General Gambiez, who was alerted, phoned General Saint-Hillier, the paratroopers' division commander, telling him to phone Major de Saint-Marc, commander of the 1st Foreign Legion Regiment, and ask what in the devil was going on. Mme. de Saint-Marc at first said that her husband was sick with a headache, which Saint-Hillier knew was not true, since the two men had messed together, and then said that he was not at home anyhow. Saint-Hillier then recklessly put the question to her direct: *"Dites-moi,* Madame, is your husband up to some dirty trick tonight?" Feebly she answered, "I fear so." Saint-Hillier and Gambiez started off at breakneck speed in their cars toward Zéralda, about twenty-five kilometres distant. En route, they saw the trucks coming toward them, pulled to one side, leaped out, stood in the middle of the road, wigwagged their arms for the first truck to stop—and jumped for their lives to keep from being run over, as some Legionnaire shouted down, "Squash the old fools!" The two generals turned their cars around, caught up with the trucks, and passed them. None of the *paras* fired on them; what the two generals did not yet know was that this was a military coup planned so as not to have a shot fired. They beat the trucks to the General Delegation Building—the seat of the city's civilian and military government—which in any struggle is the goal to capture at once. The building's forecourt has an iron grille fence around it, with two gates, both locked at night. One general stood before each gate as the parachutists arrived. Gambiez is a short-statured general; a couple of big paratroopers picked him up—their commander-in-chief—as if he were a little boy, and set him down out of the way, and officers arrested him. Other *paras* climbed over the iron fence like monkeys in leopard-spotted uniforms and unlocked the gates, and they all swarmed inside. A little later, General Gambiez was taken to General Challe, and so found out who was running the insurrection. Challe politely asked his commander-in-chief to come over to the insurgents' side and take charge, which Gambiez indignantly refused to do, so he was taken to prison—the second time that night he had been put out of the way. Some Algérois in nightclothes stuck their heads out of their windows to see what was going on. But nothing was going on, according to their standards, for not a shot was being fired, so they went back to bed. "It was a textbook *coup d'état,* a model of how to take over a city," one officer said later, adding, "The insurrection show was run by about fifty Army men"—out of four hundred thousand Army men in all Algeria.

The insurrection failed because General Challe erred in thinking that a big slice of the armed forces would defect to him. He did not get the Navy or even the Armée de l'Air; all he had was perhaps eight thousand *paras,* and he had to use them like messenger boys, sending them on first one job, then another, trying either to persuade or to bully. Challe captured and controlled buildings in Algiers,

Oran, and Constantine, but he never won whole cities, let alone all of Algeria. He never really had a base of operations. For another thing, he had not informed any of the Algerian activist civilians of the insurrection, saying that he wanted no politicians and townspeople underfoot to make another failure, as they had with their barricades insurrection of January, 1960, when they held out nearly a week in the center of the city and then let their insurrection be broken up at long distance by de Gaulle's threats and arguments. In his mutiny, Challe did not even aim at the quick overthrow of de Gaulle's government, it is said, for he had no political substitutes, or even candidates.

One young French paratrooper said, "I would not have dared to fly over Paris. My mother lives there, and if she had laid eyes on me, she certainly would have slapped my face, *bon Dieu*." Actually, it is now believed that Challe probably had no plan to drop *paras* on Paris. This will no doubt be his defense when he stands trial. The objective of the insurrection was to take over Algeria completely and at once, to liquidate utterly the F.L.N., or rebel native army, and then to turn to France and say, "Here is your liberated Algeria—liberated from the F.L.N. Here is French Algeria, for France. Put Evian and the peace negotiations out of your mind; they are already forgotten." If this sounds mad to the point of dementia, it can be better comprehended (or so the pair of visitors to Algeria intelligently explained) by taking account of the mental decadence in the diehard officer class—isolated by their six-and-a-half-year war in Algeria, cut off from contemporary, fast-moving colonial history all over the world by their morbid patriotism and their hypnotic obsession with the cult of triumph after France's long list of defeats, beginning in June, 1940. That the insurrection was to thwart the menace of Communism in a liberated Algeria that had no association with France was apparently a very secondary consideration in Challe's notion for the insurrection—though it was certainly of primary importance to the right-wing, often rich civilian accomplices that the mutiny clearly had in Paris and Metropolitan France.

It is said that Challe knew he had lost on Monday night. De Gaulle's Sunday speech had been picked up by lots of the soldiery on their transistors, as well as Debré's Sunday-night warning that the parachutists were coming. Yet Challe gave no orders for the planes to fly to Paris. It is also said in Algiers insurgent Army circles that the insurrection failed because it was an old man's revolution—because it was led by four old generals, three retired because of age. Actually, it was really a colonels' insurrection, it is now known, originated by them and then passed over to the protection of the generals, whose rank and prestige might give it the high hierarchical authority felt necessary to rally the rest of the Army. It is claimed that the idea was first conceived by Colonel Jean Gardes, former chief of the Algiers Army's Fifth Bureau of Psychological Warfare, while he was sitting in the prisoners' box in court in Paris at the trial of the barricades rebels, of whom he was one.

May 15

L'Institut Français d'Opinion Publique has taken a poll of how French citizens feel about various facets of the April French Army insurrection in Algeria, now that it is apparently over and people here in Metropolitan France can steady their thoughts. The poll presented perspicacious questions that covered a great deal of ground, and brought forth some unexpected and illuminating answers. This sampling of public opinion, printed in *France-Soir,* seems especially informative when set against the mixed confusion and flat relief that immediately followed the events. To the opening question—What elements caused the check of the Algerian revolt?—the action and influence of President de Gaulle rated top among the answers, though at only thirty-nine per cent; the determined hostile attitude of the French people against the insurgents was next, with sixteen per cent; the firm attitude against the rebellion of the French Army conscripts in Algeria drew fourteen per cent; and the attitude of the American government got one per cent. This last was an agreeable and surprising antidote to the current rumor—denied here by Ambassador Gavin—that agents of our C.I.A. had tried not to check but to cheer on the insurrection as a laudable effort to prevent an eventually liberated Algeria from possibly going Communist. To the question as to what fate should be meted out to the insurgent ex-generals, thirty per cent were for the death penalty; twenty-five per cent proposed prison, presumably for life; twenty-one per cent judiciously favored leaving them to the "extreme rigors of the law," as de Gaulle promised; one per cent thought that they should be exiled; six per cent thought that they should be shown clemency; and one per cent thought that the generals were right in what they did, and within their rights to do it, too. It is taken for granted that the last three opinions, totalling eight per cent, came from activist sympathizers, of whom only one per cent still had the spunk to speak up openly. As for the French government's immediate prosecution of those who rebelled against it, forty-nine per cent thought that it should prosecute very rigorously, twenty-nine per cent thought that mere rigorous prosecution was enough, and twelve per cent thought that the government should be clement—again that soft, sympathetic note. The answers to a question about the possibility of another Algerian uprising were in part fatalistic and alarming, with twenty-four per cent—almost a fourth of those polled—frankly saying yes, they did think one possible; forty per cent saying no; and thirty-three per cent still so troubled by the uprising itself and by the sullen, angry, anti-France reaction of the Algérois white French settlers that they said they did not know what to think. Even more tragic were the answers to the question of whether the French felt they could now trust the loyalty of their regular French Army, with which since Napoleon's time the French nation has carried on a kind of love affair. Only twenty-four per cent declared that they were very confident they could trust it; forty-three per cent felt rather confident; eleven per cent had no confidence in it whatever; and twenty-two per cent, or more than a fifth, were unable to make up their minds—signs of tepid faith in the worthiness of France's Army, which is also now gravely worrying NATO and France's allies. As for Algeria's future—which is, after all, what the insurrection was about—fifty per cent thought Algeria would ultimately be independent and would follow de Gaulle's warning and his hope that it would maintain an association with France thereafter, and nineteen per cent thought it would be independent but would try to have the remotest possible relations with France (in other words, almost three-quarters of these French think that Algerian independence is a sure thing), while only three per cent said they believed Algeria to be French property. Whether so small a proportion with the courage of its convictions covertly represents a much bigger French faction, numbed by discretion or real fear right now, when the government's dragnet is spread to pull in activist sympathizers, nobody knows, of course. Unfortunately, it is fairly easy to suppose that many of those queried on Algérie Française were simply lying in their answers. Certainly the complicity with the insurrection in high military and government circles and the sympathy for it among the rich industrial section of French society have been amply notorious. However, it is thought possible that the rapid collapse and the unpopularity of the insurrection may have brought many of France's diehards up to date at last on the futility of their fetish belief in the continuation of Algerian colonialism. On this point, de Gaulle a few nights ago, in his latest speech to the nation, specifically begged them to give up "their outdated myths."

The next-to-last question in the poll obtained the expected proportion in its major response, but also some odd by-products. It asked, "What do you think is the most important problem for France right now?" "Peace in Algeria," seventy-eight per cent said, with two per cent particularizing, "Peace negotiations with the rebel Algerian F.L.N. Army," which now seem actually set for Evian later this week. However, five per cent said that the most important problem in France now was its low salaries and standard of living. A more patriotic four per cent thought that "the stability of the regime and its institutions" was dominant. Three per cent—a low fraction of idealists—gave "world peace" as their answer. And only two per cent of the French who were questioned thought that the ex-generals' Algerian insurrection was still of paramount importance, even though it shook the French nation and them and the Western world only three weeks ago. For another two per cent, "a variety of things"—none specified—were of supreme importance to France in this still quasi-troubled hour of its history. One wonders what on earth they were. The final question was a natural: "Are you confident that General de Gaulle can settle the Algerian problem?" "Very confident," said forty-six per cent; "Rather confident," said thirty-eight per cent; "Rather unconfident," five per cent reported; and four per cent bluntly declared, "No confidence at all." It is felt that peace absolutely must come now, or the unsuccessful, crazed insurrection of military hotheads will, after all, have been victorious. In any case, the answers to the poll's final question sum up a decimal portrait of a France in a remarkably high state of unity after such a grave disturbance—indeed, in a high state of unity for any time. For the essential French tragedy is that Frenchmen always think with greater disparity than men in any other country. Yet after three years of him, General de Gaulle is still all they have—the sole great figure, with a voice, on the landscape.

As the London *Sunday Times* succinctly said in its position of a worried neighbor of France, "It is more than ever a sobering thought that the destinies of a great nation at the center of the world should depend upon the heartbeat of a single man."

November 16

During the last fortnight, if you were reading in bed around midnight in the midtown section of Paris, with the window already open and the wind in the right direction, you could have plainly heard at least some of the recent nocturnal bombs exploded by the O.A.S. terrorists. Eight were set off a week ago last Wednesday. The night before that, this writer heard, shortly after eleven o'clock, a bomb in the Rue de Ponthieu, next to the Champs-Elysées, and another up in the Avenue Franklin Roosevelt, which splintered the porte-cochere and all the windows in an apartment house inhabited by M. de Beaumarchais, an assistant to de Gaulle's Foreign Minister. If the bombings are properly planned, they are aimed at somebody the O.A.S. regards as its enemy. (Actually, many of the Paris bombings have been aimless, as if they were committed only for the sake of the noise, and to scare the wits out of the public generally.) On Monday afternoon of this week, an apartment house near the Sorbonne, inhabited by, among others, a distinguished mathematics professor who is Jewish and who had denounced racism, was so badly wrecked by a *plastic* that his library and the keys of his piano were blown into the street, along with the staircase, and firemen had to take everybody out on ladders. That same afternoon, the bombers struck the apartment building where the nationally known radio court reporter Frédéric Pottecher lives, after sending him a letter that warned him to choose either "the coffin or the valise," meaning "Leave town." Earlier, they had bombed the quarters of the *doyenne* of court journalists, Mme. Madeleine Jacob, of the morning paper *Libération*. Both of these victims had reported unflatteringly on the prisoners in the Barricade Trial. Even the suburban Château de Louveciennes, which belongs to the Comte de Paris—pretender to the French throne, and a good Gaullist—had its bomb. So far this year, a hundred and ninety-one plastic bombs have been set off in Paris and a hundred and sixty-one in the provinces.

The heaviest plastic charge yet exploded wrecked the super-popular establishment called Drugstore, at the top of the Champs-

Elysées, shortly before five o'clock this morning. The scent of perfume rising from its smashed stocks still dominated the sidewalk outside at noon, when an enormous crowd of young French gathered to mourn. The Drugstore had been the most vital center of Americanization for them, for it was a complete replica of what an American drugstore is and means in the American way of life, and had more influence on Paris youth than any American book translated into French, or any Hollywood film ever shown here.

Sixty cadavers of Algerians have been fished from the Seine or gathered from nearby wasteland since October 17th, when, at dusk, thousands of Algerians living here surged out of the Métro stations into central Paris in protest against and in defiance of an eight-thirty curfew. On Tuesday of this week, there was a painful discussion in Parliament about the bloody brutality of the Paris police, who were held responsible for these deaths, though it was admitted that gang warfare between rival Algerian groups might account for a few corpses. The new appropriations for the Department of Justice were also up for a vote on Tuesday, which was awkward. Furthermore, three deputies who had been officially sent to inspect the Vincennes reception barracks, where thousands of the anti-curfew Algerians are still detained—many with medically unattended head wounds, alleged to have been caused by police clubbings—had reported that the conditions there were "scandalous." The Parliamentary discussion about the police and the Department of Justice was thus enlivened by passion and political venom, which added to the logical difficulties of Prime Minister Debré, who was presiding. The Paris Prefect of Police has declared he will sue any newspaper that criticizes his policemen—the only funny item so far connected with his gendarmerie's October 17th mayhem and slaughter.

The tortures practiced in Algeria by the Fifth Republic's Army—so horrifying at first to civilians—at the time when the French were still fighting to keep Algeria a French province have disappeared from the news now, and the increasing Fifth Republic noise of exploding plastic bombs in Paris seems neither to distress nor to impress Parisians—unless, of course, their apartment is one of those wrecked—for a certain degree of violence, of terrorism, and of cruelty appears to be a normal part of present-day French life. Both the torturing and the bombings were from the beginning repeatedly deplored in printed manifestoes by certain intellectuals, but their energy has gradually faded away, like ink. It is an accepted fact that the O.A.S. and its bombs represent the sentiments of—and are paid for by—the French Rightists, which means, socially, some of the nicest people in town, well educated, well-to-do, and *comme il faut.* The bang of their bombs going off furnishes a peculiar, brutal sonority in Paris, once considered the most civilized and cultured spot on earth.

December 5

There was a unique domestic quality about Miss Gertrude Stein's wonderful collection of modern French pictures, because they were a major part of her household. Over the years, they kept company with Miss Stein and her friend, Miss Alice B. Toklas, and the ladies, on their side, maintained a close companionship with the canvases, in the civilized intimacy that relates certain human beings and objects when they have long lived together. It was a collection with its own kind of private life. It was the oldest permanent collection of modern French art in Paris, because it was formed in appreciation of what no one else then wanted, and certainly it was the only one that, while illustrating new forms of Ecole de Paris art on the Stein walls, was also a literary witness to the creation of the new school of American writing in the Paris twenties, which first took shape in endless, important talk—often animated by young Hemingway's strong voice—in the ladies' salon. When Miss Stein suddenly died, in 1946, and her collection was willed to Miss Toklas, "to her use for her life," the inventory listed twenty-eight Picasso pictures, one Picasso sculpture of a head, twenty-eight Picasso drawings, and seven Juan Gris canvases. In late April, 1961, the collection was removed from the Stein-Toklas apartment, in the Rue Christine, on the order of the Tribunal de Grande Instance de la Seine, and it is at present in the vault of a bank in Paris under the trusteeship of a court-appointed administrator. Today, the only trace of these masterpieces in the apartment is the faint, empty outlines left by the picture frames on the white walls. Miss Toklas—now eighty-four and fragile, though spirited—was fortunately not present for the shock of seeing Miss Stein's pictures borne away. She had been absent in Rome—precisely the opening basis for the Tribunal's act—ever since the previous autumn, so as to avoid the chill of the French winter and late spring. ("My first infidelity to Paris, and a big mistake," she says succinctly.) Only her close friends have known of the immolation of the Stein paintings, though not what lay behind it—of which even Miss Toklas, alas, at first had no inkling.

It seems that, on sound tax counsel, Miss Stein described herself in her will as an American who was resident in Paris but legally domiciled in Baltimore (a sort of vague second home for her, where she had attended Johns Hopkins as a student and still had relatives), because American death duties start only on a net estate exceeding sixty thousand dollars, whereas the French taxation starts with the first franc if the inheritors are not in direct family line. In the case of a collection that had become as valuable as hers, the high French duties could have meant the enforced selling of some of the contents so that Miss Toklas might inherit the rest. The will named only two reasons for which any pictures might be "reduced to cash." The first was to permit the publication of any remaining unprinted Gertrude Stein manuscripts—all willed to the library of Yale University, which subsequently published eight volumes of them. The second was to provide for Miss Toklas's "proper maintenance and support," if necessary. However, any sale had to be authorized in Baltimore by a court-appointed estate administrator (who, early on, was an elderly lawyer named Edgar Allan Poe, the poet's great-nephew, who died last week). This authorization Miss Toklas seems to have had in 1958, when she sold a Picasso "Paysage Vert" to M. Daniel-Henry Kahnweiler, the venerable Cubist dealer and a friend from Cubist days, for $18,750. In 1953—without Baltimore authorization but for one or both of the reasons stipulated by Miss Stein's will—she had sold Kahnweiler, for about $6,000, half of the twenty-eight Picasso drawings, later exhibited and put on sale at the Galerie Berggruen, in the Rue de l'Université. The prices were modestly set by Picasso himself, Miss Toklas says, whom she had asked for advice. Fairly recently, one of these Picasso drawings was resold at Sotheby's, in London, for three thousand pounds.

Late last year, the British news came to the knowledge of the widowed Mrs. Allan Stein, here in Paris, and the recent drama started, Mrs. Stein's interest in art being strictly that of a mother. Her late husband, Allan, the son of Gertrude's brother Michael, will be known to art experts through an early Picasso portrait of a little boy holding a tennis racket, though his adult sports preference was reportedly race horses until his early death a few years ago. It was to nephew Allan that Miss Stein arranged for her art to go after Miss Toklas's lifetime, and now, with his death, the new heirs will be his eldest son by a first marriage—today in his thirties and living in California, where Miss Stein grew up—and another son, Michael, and a daughter, Gabrielle, both in their young twenties, who are resident in Paris with their mother, who regards herself as their art guardian. As such, Mrs. Stein early this year demanded a new inventory of the collection they will ultimately inherit, and this revealed the absence of the twenty-eight Picasso drawings, which, though they had been sold for reasons that Miss Stein's will approved of—half were sold to pay for publishing the last of Miss Stein's manuscripts—put Miss Toklas technically in the wrong with both French and American law. The action brought by Mrs. Stein opened with a complaint against Miss Toklas's protracted absence in Rome, which left no one living in the apartment with the pictures, and went on to complain that even when she was in residence here she was alone and defenseless at night amid such great art, that there were no bars on the apartment's windows against burglars, and a few other items. All the complaints were concentrated in a demand to the Tribunal that, for the good, the protection, and the preservation of Gertrude Stein's collection—today worth more than a million dollars, with an estimated value of $150,000 alone for the great rose nude on a gold background, unlike any other Rose Period picture Picasso ever painted, which used to hang in the Stein salon—all the pictures be immediately put in a safe, dry, guarded place, thus landing them in the Chase bank in Paris.

Miss Toklas's eyesight is, naturally, not what it once was. Of the disappearance of Miss Stein's familiar pictures from her salon and foyer, she only says, "I am not unhappy about it. I remember them better than I could see them now."

December 13

In many ways, this has been a troubling, disputatious, and disillusioning year for the citizens of France, and also for plenty of others—white, black, and yellow—scattered around the world, worried or hungry or blood-stained in villages or jungles, or cut off from their neighbors by a new city wall, or harried by arguments and democratic perplexities in the high seats of government, or buffeted by harsh international debates in a certain glass skyscraper, with a fifty-megaton explosion in the East and plastic bombs in French doorways having become more vivid to most people than the Star of Bethlehem at this season. No matter what paper you pick up in Paris, it contains its budget of painful news of some sort. One paper declares that France is living on the edge of chaos. One editor says that the civil war that has ravaged Algeria for seven years has, by terrorism and counter-terrorism, spread to the mainland of France itself. One evening journal's analysis concludes that an ever-widening section of the French population is "ceasing, little by little, to live within the law." Straight news consists of reports on the thirty or more inhabitants of Oran or Algiers who were bombed to fragments in cafés over the weekend, or on the more or less daily plastic explosions, so far luckily without deaths, in Paris, or on the leftist protest marches against these rightist bombers all over France—in Toulouse, Rouen, and Lyon—with fracases and broken heads in the struggles between the marchers and the police. Tuesday night, thirty bombs were set off in Paris, of which eleven, with good luck, failed to explode. All were accompanied by tracts of the Organisation de l'Armée Secrète, which the explosion was supposed to scatter. Of those that went off properly, two were on the Champs-Elysées (where five were duds), and single ones were in the Place de la Sorbonne, the Mairie of the Sixième Arrondissement, not far from Place St.-Germain-des-Prés, in a church, and in the Gare du Nord; two bombs were in the Gare Montparnasse, and four in the Gare de l'Est.

It may seem astonishing (though, after all, it is only logical) that there is a whole section of the Paris press—on public sale, like any other papers and periodicals—that serves the many readers sympathetic to the extremists of the O.A.S., without, of course, openly cheering their plastic explosions, which it usually refers to as "of unknown origin" or as the work of Algerian terrorists, or simply does not report. The two dailies of this section of the press are *L'Aurore* and—much more powerful—*Le Parisien Libéré,* with a circulation of eight hundred and fifty thousand. The latter also prints the weekly *Carrefour,* with its tendentious editorials and gossip—a journal that recently described itself as "in no way a Fascist enterprise but a force of resistance to oppression," meaning chiefly the government's policy of Algerian self-determination, which the *ultras* still resist as a plan to rob France of part of herself. The weekly *Aspects de la France* is also pro-*ultra* and anti-Fifth Republic. All are anti-Communist in a defamatory manner, rather like our John Birch Society. The weekly *Rivarol* is the boldest and most insolently anti-de Gaulle. Lately, it declared, "De Gaulle can puff himself up all over France, but it is only the ignorant or the imbecilic who believe that he is qualified to preach about obedience."

The biggest recent press event was a large plastic bomb set off last week in the editorial offices of *France-Soir,* which fortunately hurt no one seriously and did an incalculable amount of good, because this popular apolitical afternoon *journal d'information* has a circulation of well over a million and a readership of five million—the largest in France—and is sold daily by its own special venders on town streets literally all over the country. News of the attack came like a personal outrage to its myriad readers everywhere, to judge by the fantastic volume of letters that began pouring in to the editors, proving a sense of personal identification or of shock, at last, instead of the curious apathy to violence that has spread like a miasma over much of France. In a special political editorial, *France-Soir* declared, "The French population has a right to be protected. It has equally the right to be informed about the measures taken against the *plastiqueurs* and those who animate them, as well as about the results of accusations brought against them. [This was a polite way of pointing out that, despite repeated demands from the public, the government had taken no such measures at all.] Public order must be maintained by those who are in charge. Justice must be rendered by those whose mission it is. The authority of the state must be imposed by those who have received its mandate. It is for the press to clarify public opinion." And it ended firmly, "Nothing will prevent it from doing its duty." This indignant professional determination was, of course, natural, and, indeed, it had already been manifested. *France Observateur, L'Humanité,* and *Le Monde* had already been bombed once each, and the apartment of the editor-in-chief of *Figaro* had been bombed twice, without any subsequent soft-pedalling of their anti-O.A.S. reporting or their criticism of the government's inexplicable inertia in taking legal steps against these crimes. The inertia was finally dispelled last Friday, when the *Journal Officiel* printed the government decree pronouncing the dissolution of the Organisation de l'Armée Secrète, with heavy fines and fairly light prison sentences for those found guilty of membership, of serving as accomplices, or of giving funds to it. This last will make trouble for many banks, oil companies, and big business, which have been blackmailed into O.A.S. financial contributions so as to be left in peace by the bombing squad. The decree was accompanied by orders for the arrest, on a charge of plotting against the authority of the state, of ex-General Salan and ten of his O.A.S. chiefs—provided anyone can lay hands on them, since they have all been successfully in flight or in hiding since July, when the Paris military court sentenced most of them to death for their part in the spring Algiers insurrection against the French Army. On Friday afternoon, shortly after the dissolution order was made known here, the O.A.S. impudently hoisted its black-and-yellow commando flag—and not once but thrice—on the roof of the Hôtel de Ville, right over the heads of the City Councillors, who were meeting there.

Yesterday, the enfeebled Socialist Party (which has just announced the suspension, at the end of this year, of its pitiful one-page party newspaper, *Le Populaire,* an influential, brilliant daily back in Léon Blum's day) brought a motion of censure in Parliament against the Debré government, which "by its maladdress, weakness, and inner divisions has lost the authority necessary to meet the threats accumulating against the Republic." Naturally, the Socialists do not expect to obtain anything like the majority of votes needed to make Debré fall, for the Chamber is not of the brilliance or the unity required even to imagine any premature uprising against what is really the personal government of General de Gaulle, who is still France's single necessity and only hope for eventual peace in Algeria. The censure motion, which will be debated Friday, appears to have been merely a straw in the wind for the deputies to let blow by before Parliament closes down for Christmas, not to open again until the spring of 1962.

The widest dissemination of personal glory for an old great artist who is still alive must be to have one of his paintings become a postage stamp—to have the vast postal system of his country become, in part, his art gallery, to have an envelope with a mere dull business letter inside, or even an airmail postcard, serve as a small, inexpensive travelling exhibit of his genius. This is the glory that has come to Georges Braque, now in his eightieth year, who was honored this last week by the French postal authorities with the issuance of a very large, very pretty fifty-centime stamp showing the long-necked white bird that has rather become his trademark in his old age silhouetted in flight against some dark-gray foliage on a pale-blue background. Braque was the literal creator of Cubism, the most famous, influential, arbitrary invention of the early *école de Paris*. So why, art lovers here are wondering, could the Braque stamp not have reproduced, for instance, his "Still-Life with Musical Instruments" of 1908, which any village postmistress today could easily identify as a Cubist horn, concertina, and lute? Or, best of all, why not have made the Braque stamp a full-blown 1910 example of his arcane analytic Cubism, such as "Woman with a Mandolin," showing music visibly recumbent upon geometry—an art work that would have perfectly signalled his greatest period and established a bold aesthetic exception even for French modern philately? France is now the first country to start using its twentieth-century artists' paintings for postage designs, and has, for beauty's sake, installed new presses that print six colors in a process believed to be superior to anybody else's.

In the fifteenth Paris Salon de Philatélie last week were shown two other notable new stamps honoring modern masters, though now dead—a stunning sixty-five-centime Matisse, bearing two of his cut-out blue female nudes (all these art stamps are so big—two inches by an inch and a half—that the Matisse heirs decided that a nude pair was aesthetically necessary), and an eighty-five-centime Cézanne, of "The Card Players." This is the painting that was lately stolen from the exhibition in Aix-en-Provence. Of the three stamps, the Cézanne is obviously the public's favorite—two provincial Frenchmen swigging wine and playing *belote* in a *bistro*. Its dark background makes it a less satisfactory example of graphic coloring, the stamp collectors say—probably an opinion of minimum consequence, for it is the picture itself that counts, the lost Cézanne masterpiece with the two men familiarly playing their card game on a postage stamp.

December 27

The sporadic feverishness in the history of the Fifth Republic seemed well diagnosed last week by an official spokesman when he said that the government refused to choose between "the plague and cholera," meaning that France is menaced by Communism as well as by Fascism, those opposite political maladies, and that one is no better than the other for the health of the nation. He was referring specifically to the enormous unified protest marches held all over France at six o'clock on the Tuesday afternoon before Christmas, which were organized by the Communist-led Left against the Rightist Organisation de l'Armée Secrète and its planters of plastic bombs. By the time he spoke, which was on Wednesday, the march here in Paris had already resulted in about thirty hospitalizations from among the hundred or more injured participants—half of them marchers, half police—in a bloody two-hour struggle on the streets between the Place de la Bastille and the Hôtel de Ville. Since the marchers had been, for once, glad to demonstrate support of de Gaulle's government, because of its belatedly declared fight against the O.A.S., they seemed to suppose that the government would—or anyhow should—back them up in their demonstration, and so did the editor of *Le Monde,* who rarely makes so optimistic an error. Why, he asked, in a short, disillusioned editorial entitled "Contre un Néo-Nazisme Français," had not all the authorities (civil, military, spiritual) planned to direct and utilize Tuesday's popular demonstration in defense of the Republic, since the Republic had itself—through the mouth of Premier Michel Debré, speaking over the radio—appealed to the public for help in those dark, frantic midnight hours of the April *Putsch,* when Debré thought the insurgent French parachutists were about to drop onto France? "Must we go through another such nocturnal tragedy, with the alarmed French people suddenly summoned to rush forth without arms or support, in the middle of the night, on foot, in their cars, or on horseback?" *Le Monde* gibed bitingly. What is generally regarded here as the government's lack of political psychology was further demonstrated by its disdaining to recognize that the major police unions had earlier sent letters to President de Gaulle, to Debré, to the Ministry of the Interior, which controls all the police forces, and to the prefect of the Paris police, protesting against the government's ban on the anti-O.A.S. demonstration—a ban that raised a "question of conscience" for those of the *gendarmerie* who were also against the Rightists but would have to fight the Tuesday marchers instead. This they did, once they got started on the job, with exceptional professional brutality—as, apparently, the prefect had ordered them to "draw blood."

The newspaper photographs the next day were appalling, showing a melee, at close range, of hatless civilians trying to hold one hand in protection over their skulls—smashed hands and cracked heads were the commonest hospitalized injuries—while the police flailed, as if during a harvest, with their extra-long white riot sticks and the helmeted special security troops used their rifle butts as truncheons, pounding with them, like pile drivers, into the massed marchers, or into the backs of men who had fallen to the sidewalk. Four marching municipal councillors, three of them Socialist and one Communist, wearing their tricolor sashes of office, were beaten, like anybody else, and injured—one on a hand, two on the head, and one in an eye. A third of the marchers hospitalized were women, knocked down and trampled on in their incredibly determined participation in the hurly-burly. Among the uninjured female marchers were Mme. Simone de Beauvoir, who walked with Jean-Paul Sartre; Mme. Jeannette Vermeersch, wife of the Communist Party leader Maurice Thorez; and the Communist woman mayor of Bobigny, a Red suburb.

The banners the marchers carried mostly bore the slogan *"Contre l'O.A.S., Pour la Paix en Algérie p[illegible] la Négociation,"* but what the marchers mostly shouted was *"L'O.A.S., assassins!"* In the circumstances, it was natural that on Wednesday morning the Communist *Humanité's* front page should be a mixture of falsehood and jubilation. An enormous headline declared, *"Plus de Cent Mille Manifestants dans les Rues de Paris."* The police figures—also probably false—put the crowd not at a hundred thousand but only at fifteen thousand, and most newspaper reporters settled for about twenty-five thousand. But thanks to the marchers' resolute facing up to their inevitable beating by the police, the Communists emerged with an invaluable new front-page slogan, which at this moment in the history of the Fifth Republic seems to many non-Communist ears to ring true: "The people of France can count only on their own force to defeat Fascism." On Wednesday, the Fifth Republic was also inevitably accused, by the organizers of the march, of *"de-facto* complicity with the Rightist activists." In the lead among the organizers had been France's most powerful labor union, the pro-Communist Confédération Générale du Travail, supported by the Leftist teachers' and students' national unions, and also by the non-Communist Confédération Française des Travailleurs Chrétiens. The Socialist labor union, Force Ouvrière, had refused to take part, because it won't touch anything the Communists have a finger in, but the new Socialist splinter party, the Parti Socialiste Unifié, to which M. Pierre Mendès-France belongs, furnished some marchers on Tuesday and, on Wednesday, a piece of its mind. It said that the de Gaulle government had mobilized more police in one evening against the anti-O.A.S. defenders of the Republic than it had "in months against the Fascist killers themselves."

President de Gaulle is to address the nation on television this Friday. It is a nation in which Left and Right are once more openly and dangerously at loggerheads, because, in a way, the French Revolution of 1789 has never been finished.

The winter art season opened late, coming to its full growth only close to the holidays. The major one-man show of the season was the retrospective exhibition of more than three hundred paintings by Mark Tobey at the Musée des Arts Décoratifs—the first time an American artist has been so honored there. It is the greatest honor Tobey, who has just passed his seventy-first birthday, has ever been given, and greater by far than any he has been shown in the land of his birth. The fact that this museum is physically a part of the Louvre and spiritually a semi-official focussing point for contemporary artists of the highest rank—Picasso was the first to be given a retrospective there, and last year Chagall had one—adds to the kudos that Tobey has won here. He first became significantly known in Paris only a few years ago, when, as he says, he and his paintings were generously hailed by Georges Mathieu, head of the so-called French Tachiste school, whose followers regarded Tobey not only as an exponent of their style but as a forerunner—in fact, a kind of unconscious founder—of it. This seemed to Tobey not to be true, without diminishing his pleasure in so friendly a brotherhood. The appreciation he has aroused among the French public and critics is like a blossoming in winter. As one critic said, he is "one of the most singular artists of our time, but little recognized in his own land." The critic went on to declare that Tobey "could be provisionally described as the painter who introduced the Oriental spirit into nonfigurative contemporary art." Because of his invention of his "astonishing white writing," he was further credited by this critic with having left behind "the forms of expression that founded modern painting," and with having, by his "fundamental originality, invented a new space, completely abolishing volume, to the benefit of modulation and rhythm." The critic of *France Observateur* said of

Tobey's white-writing style that it was "a discovery on the highest plane of present-day art; a momentous entry into sensibility and silence," as opposed to the clamorous quality of much of the painting done now. The magazine *Preuves* called him "perhaps the most important painter of our epoch," and went on to say, "In Picasso and Tobey, who is only ten years younger, the future lies prefigured. Tobey's labyrinths, trellises of color, nebulae, and eruptions of atoms are, in their true signification, the measureless universe."

Tobey at present is back in his seventeenth-century house in Basel, where, as if his paintings were enamels, he is once again slowly drying them in the vast cooking oven of his kitchen.

1962

January 9

It takes at least a week to get a couple of tickets to the little Montmartre Théâtre de Dix-Heures, owing to the popularity of Henri Tisot in his now nationally famous and uproariously funny parody of "qui vous savez"—as the French regularly identify his subject, their not openly mentioning General de Gaulle's name somehow adding to their irreverent glee. What Tisot has created and gives is a brilliant vocal and intellectual takeoff of de Gaulle the speechmaker—a total mimicry of the august voice, with its sometimes uncontrollable comic falsetto and its solemn nasal trumpeting of sacred syllables like *"la Fraaaance,"* coupled with bull's-eye precision in pinning down the *démodé* verbal elegances of his vocabulary. Most astute of all is the imitation of the General's style of special thinking—of his complicatedly pointing out or splendidly obscuring his ideas, which has become his individual rhetorical pattern in historic oratory. The takeoff is based on those de Gaulle speeches to his nation that referred to auto-determination for Algeria, which Tisot has turned into *"autocirculation,"* or the terrible traffic problem in Paris. A recent Pathé record of the parody, made live at the Dix-Heures with the audience's roars of laughter, has in the last month sold almost a quarter of a million copies—a hit not equalled in France since the "Third Man" zither tune. Tisot, who is thirty-one, has something of the heavy facial structure of the President himself; can purse his lips like the Chief of State for similar labial effects; uses his hands, thumbs up, exactly as the General does; had his training at the Comédie-Française, where he was rarely given a chance to act; and is now nightly doing his *"autocirculation"* parody in a special early turn at the big Olympia Music Hall, where, to add realism for the large, non-intellectual audience, he stands on a speaker's platform draped in official tricolor bunting, which is rather daring.

As a fact of de Gaulle's Fifth Republic's oratorical history, his New Year's radio speech was the most unpopular he has yet made to the nation, with its lithographic, optimistic portrait of France as an island of stability, self-confidence, and prosperity amid a world in tumult. His speech to listening millions sounded as if he were uninformed about the opposite present truths, and is still provoking caustic references. In a recent *Figaro Littéraire,* to whose back page François Mauriac has transferred his trenchant "Bloc-Notes," he comes forth as de Gaulle's sole apologist. He explains his belief that the reason de Gaulle said nothing consequential to the French people at New Year's was that he knew too much to be able to say anything with safety. Mauriac states, in part, "Never has assassination that has become endemic in a country coincided with such languor. We hate and we kill meanly and apathetically. It was not for de Gaulle to tell us what we have become. It was not for him to hold the mirror up to us. As always at the bedside of someone very ill, nothing was left for him to do but to voice words of hope. He could not cry out to us *'Eh bien, oui,* Frenchmen, you are in a bad way and no pleasant sight to see.' This he could not say." Unfortunately, many Frenchmen still think that this is exactly what he should have said, and loudly, too.

The incredible price of fresh black truffles over the holiday season—the highest ever recorded in modern French gastronomy, and coming, of course, just when they were most needed for slipping under the skin of the turkey's breast or into the goose's liver—seems to have aroused truffle raisers to an almost confessional frame of mind. An authoritative article on trufficulture, compiled by government mycologists, biologists, truffle growers, and barren-soil experts, has, as a rarity, just appeared in a weekly farm newspaper, *La France Agricole.* It makes confused, fascinating, and mysterious reading. Truffles, which for years had been selling quietly at the equivalent of seven or eight dollars a pound, in December suddenly leaped to twenty-one dollars, or about two dollars and thirty-five cents for one warty, thick-skinned, coal-black, high-scented, delicious *Tuber melanosporum* the size of a Congo baby's fist. This peak price both delighted and alarmed *trufficulteurs,* whose present drama it indicated—"the penury of growing truffles, which are extremely capricious in their reproduction" and "have an evolutionary cycle of which little is known." What is very well known is that Italian and Hungarian black truffles are now competing in the international market to fill the shortage (lately amounting to two-thirds of the former crop) of the famed Périgord type of truffle, which has long furnished its erratic troves of subsoil riches from among the thousands of acres of otherwise unredeemable stony land in the southwest, especially in Dordogne, Lot, and the Corrèze. (The country inns down there serve you sliced truffles on everything but a boiled egg.) Apparently, only one simple fact is known about a truffle, and it used to be denied. It is definitely an underground mushroom. According to the official Truffle Committee of Périgord, truffle growers perforce know little about growing them, except that if they grow at all, they mostly do it under pubescent (or hairy) oaks—though they won't do it even there except maybe under ten trees that have "a special truffle vocation," out of a hundred saplings that the wretched farmer may have planted, in clay above fissured limestone. He plants them not upright in a hole but lying sidewise in a trench running north and south, with a supporting stick to lift the branch end, so that the leaves can later give their miserable shade to the roots near which the truffles perhaps will grow. It takes the trees about eight years to reach their stunted and scrofulous-looking maturity, when they should manifest their vocation, if any. In that case, the ground around the tree "burns," as the peasants say—the lichen or sparse moss becomes mysteriously scorched. This, "we think," the Périgord Committee opines hesitantly, "proves that the parasite truffle is living in a state of symbiosis with its tree"—from then on honorably called a truffle tree, or *un chêne truffier*. "It is certain," the Committee says more firmly, "that the tree combines the unknown conditions that give the determining cause for truffles. It may be a secretion from an injured root. Truffle-making is the particular characteristic of a certain tree, not the normal function of the species." The great plantations of 1900, now dead and creating the present shortage, flourished in old vineyards that had been killed by Phylloxera (accidentally imported from California). Since the isolation of the truffle mycelium in the early nineteen-forties by a Clermont-Ferrand botany professor, it has been used to inoculate young oaks, and even their acorns, for starting new plantations. But no one seems to know if it really works. There was recently a celebrated truffle tree near Meyssac, in the Corréze, which had a truffigenous radius of a hundred and twenty square feet, and between 1954 and 1958 produced twelve hundred dollars' worth of truffles. This is why truffle growers and government agriculturists wish they knew more about the truffle's private life. Brillat-Savarin called the truffle "the black diamond of the kitchen." It is possibly a delicious, edible underground mycological malady.

January 25

The heaviest, so far, of all the plastic bombs set off in Paris exploded Monday afternoon in a courtyard of the Quai d'Orsay, and its hollow boom could be heard as far away as the Place de la République, about three kilometres distant. It was a lovely, clear-skied late afternoon, and many of us living high up in the hotels along the Rue de Rivoli had a window half open, so the alarming sound, bigger than any we had yet heard but familiar, came to us direct, across the river and the Tuileries Gardens. From our little balconies, we could see a column of smoke starting to rise from the part of the Foreign Ministry across from the Invalides air terminal, and in a few minutes we could lean down to see and hear the fire engines and ambulances plowing through the Rivoli traffic, their sirens raising a bedlam and their big red lights winking as they hurried toward the Concorde bridge. This outrage committed against this major government edifice—at first, people thought it was Parliament, next door, that had been bombed—has shaken, angered, and horrified Paris at last. On all sides one hears the French saying that things can't go on like this, why isn't someone doing something to stop it, where is the government? On Tuesday, *France-Soir* published a front-page photograph of the body of a man who looked made of charcoal, with a caption explaining that "this atrocious document" showed "the cadaver of the modest Quai d'Orsay employee" carbonized by the O.A.S. Monday bomb (miraculously, the only death from plastic bombs that Paris has had), and that although the policy of the paper was against pandering to morbid curiosity, it was high time for "public opinion to be a judge of these terrorist practices." Monday morning, a Gaullist deputy had been kidnapped by men with machine guns—"Chicago-style," as one Paris paper put it—and that night, in the new Paris style, the police searched the domiciles of four hundred people for incriminating documents and arrested twenty-nine suspects, many of them teen-agers from the best of the city's *lycées*. The activity of the police, so long inert, and the entry of decent people's young sons into the nasty arena of events have been the week's unexpected big political elements. In the Sorbonne's law-school classrooms, right-wing student sympathizers recently painted enormous "O.A.S."s on the ceilings and "Algérie Française"s on the walls. This pro-Fascist youth movement is opposed by the biggest student union, which is pro-left, their dual political precocity being a natural microcosm of the warring split among adult partisans in France.

Monday night, the homes of half a dozen Communist deputies or officials were plastiquéd; yesterday it was the liberal journalists' turn. In less than half an hour, after lunch, seven bombs partially wrecked, among other homes, the apartments of M. Beuve-Méry, the editor of *Le Monde* (it was his second plastic); of M. Maurice Duverger, a Sorbonne law professor and a frequent *Monde* political writer; of Mme. Françoise Giroud, co-editor of the leftist weekly *Express* and of Michel Droit, the editor of the weekly *Figaro Littéraire*.

Since the Quai d'Orsay Monday bombing, there has been a new, urgent, angry tone to what most of the Paris papers have to say about the Fifth Republic government and its chief. *Figaro,* which until now has been carefully courteous in expressing its dissatisfactions, lost all decorum in its editorial yesterday, and in being truthful became tough. "Yes, General de Gaulle disappoints us," *Figaro* said, in part. "Yes, the ignorance amid which he has installed himself at the top of his ivory tower, and the obstinacy of his disdain, seem in defiance of current events. Yes, the sense of attachment to him is no longer present in most hearts. But he has at least the immovable sentiment of fidelity to his duty. We have no choice, and neither has he." On this subject the satiric weekly *Le Canard Enchaîné* also had its say, in a cartoon that showed President de Gaulle in his nightshirt sleepwalking on top of a Paris building. Inside, Prime Minister Debré is leaning from a window above a crowd waving banners marked "Stop the O.A.S." and "O.A.S. Assassins," to whom he whispers, "Quiet! You will wake him."

The cleaning of that splendid row of Place de la Concorde palaces built by Louis XV for the reception of ambassadors and potentates, which in our tourist epoch have been mostly identified with the Automobile Club, the Hôtel Crillon, and, until lately, the Guaranty Trust Company, is now finished, after months of detergents, and the transfiguration is complete. You can hardly believe your eyes. The formerly dramatic blackened porticoes and pillars have become elegant cream-colored façades of delicate nuance, of carved intricacies behind faintly runnelled classic columns, all restored to their original, shallower perspectives. These buildings are the pilot portion of a grandiose clean-up program for certain notable *monuments historiques* lately undertaken by M. André Malraux, Ministre des Affaires Culturelles, who, with characteristic aesthetic flair, chose to launch it on the Concorde as "the most spectacular site." As pendent extensions of the Concorde's view, the Madeleine and Parliament will also be scrubbed this summer. The cleaning up of the Chamber of Deputies, even on the outside, will indubitably be reflected in the press by harsh political cartoons and jokes on the scandals that have too frequently stained republican history there. Also to be cleaned by autumn are the three long, richly pilastered, and fastuous garden façades of the Palais-Royal (where Louis XIV lived as a boy, and which still contains the most secluded, coveted walkup ex-royal apartments in all Paris). With its rehabilitation, the Palais-Royal will look like an enormous restored painting, its sepia-tinted stone emerging as a pale, rich chrome. This is already the color revealed in the garden's Cour d'Honneur, where Malraux has his Ministry offices, and which he ordered cleaned first, as a sample. His 1963 program will include the red brick Place des Vosges (built for Henri IV) and the opulent Place Vendôme (known to Louis XIV only when he was an old, unpopular monarch). Malraux's extraordinary effort to freshen certain great elements of the French architectural past, to redate beauty by bringing back, as far as possible, the complexion and coloration they had when their period was young and royal—or at least to repeal half a century of recent time and the black crust deposited on this beauty by modern prosperity, by industry's and motor traffic's fumes, and by today's atmospheric filth—has naturally divided Paris opinion into two camps, this being a long season of unflagging, nerve-racked French dissension and dispute about everything, about anything. Those against say that making the old buildings appear younger has robbed them of their weight of reality, that the Concorde palaces now look like Hollywood movie sets of the Concorde palaces. Among the owners of these sites, only the Hôtel Crillon was reportedly opposed to being cleaned up, on the ground that its clientele, especially the Americans, had faith in the dirt on the façade as a guarantee that they were living in a genuinely historic old place.

Apparently there have been many cleaning methods used on old stone here in France, several of which have been dropped because they were injurious either to the stone or to the men working on it. Sandblasting is now forbidden here as perilous to the workmen. Steam jets have proved too brutal for antique façades, but jets of water at high pressure are still permitted. Eighty years ago, the Germans undertook to wash the Cologne Cathedral. The English washed Buckingham Palace with water and sponges. Nowadays, the atmospheric filth of big cities is so adhesively gummed onto the old stone that detergents are necessary. According to experts, the great danger in cleaning old stone is the destruction of its *calcin,* or epidermis, whose removal "kills the stone," in the masons' phrase; to renew the stone's resistance to weather, it has to be treated with fluosilicate. A stone virus, or sickness, has also recently been discovered. The cleaning in the Malraux program is done with a detergent powder of secret and antiseptic formula, invented in Germany, which is mixed with water to the consistency of paste, is applied to the stone with a mason's trowel and left to act a few minutes, and is removed by a stiff scrubbing brush, the stone then being rinsed off with water—rather the same general process as brushing one's teeth. By a law of 1852, French property owners must clean the front of their buildings every ten years. In the case of the state's *monuments historiques,* M. Malraux has decided that the state should itself submit to the excellent law it imposes on its citizens.

February 7

This being so palpably a moment of near-final crisis in France's Algerian ordeal, de Gaulle's Monday-night speech was considered of major importance throughout the Western world as well as here. The London *Times* at once signalled it as "one of his most persuasive discourses." On the whole, the ordinary French people seemed satisfied with it, and with the last part especially, because there, for once, he was speaking factually—an exception they were hungry to hear. The Paris press is distinctly less critical and hostile than after his December 29th address, in which his ill-starred soothing phrase that "all is simple, all is clear," along with a few other paternalisms, combined to give him for the new year what could be fairly called a unanimous opposition press. But since Tuesday it is more pro-de Gaulle than it has been for a very long time indeed. It is through his words—far more than through his deeds—that he rules again. For itself, *Le Monde* rallied to him fraternally with its own special gift of language in its front-page editorial, saying, in part, "Yes, for the first time perhaps in seven years, these days we are close to our goal and, before that, to an end of the fighting. There was something moving, tragic, in seeing, in hearing this lone man"—without naming him—"intrepid, immovable, sure of himself, seeking to communicate here and elsewhere the certainties at which he has arrived across the eddies that for three years carried him so far in the opposite direction from what he had hoped, planned, declared, and to which events cruelly gave the lie at a time when he believed himself powerful enough to command them. These certainties, which he gained at the cost of his own disavowals, are that there is a determinism to history, that Continental France retains its potential as a great power, with an eminent role to play in the 'union of organized states' that will be Europe, and that no alternative exists to the phenomenon, soon to be universal, of decolonization."

As for the extreme-Right newspaper reaction to de Gaulle's speech, the full folly of it is spread in the news of the Algerian response to the speech as detailed here in Tuesday's *Aurore,* the Paris Rightists' morning paper. It said that, in obedience to O.A.S. orders, the streets of Algiers became deserted just before the broadcast; that when the speech began, a deafening tumult broke out, with the anti-de Gaulle *colons* blowing trumpets or beating drums or saucepans from their windows; that in the harbor the tugboats blew their whistles throughout the twenty minutes of the speech; that, hidden inside their parked cars at the curb, other anti-de Gaulle Europeans played "ta-ta-ta-ti-ti" on their horns, which scans "Al-gé-rie Française." In Oran, most violent of the cities, the long-wave radio transmitter had been plastiquéd in the afternoon and its director and technical assistants temporarily kidnapped. In the city of Bône there was a *"concert de casseroles,"* or frying pans, plus sixteen plastic bombs, including one that exploded in the police prefecture during the General's address; at Mostaganem, one plastic bomb per minute was set off during the speech's twenty minutes. "And that is how the discourse of the Chief of State was welcomed in the Algerian cities," *Aurore* commented. About twenty people in Oran and a half dozen in Algiers were murdered on Monday.

February 20

The only words that will satisfy the French people will be brief and printed ones, headlined on the front page of every newspaper, declaring that the cease-fire has been signed. It won't actually be peace, but it will feel like peace, because it will mean the end of the two armies' shooting at each other in Algeria, as they have been doing for seven and a half years. Maybe it will come this weekend, or the one after. It is the only event that could follow suitably on two earlier events that people here will long remember and talk of, and that will become modern legends of Paris. The first event, of course, whose importance lay in its horror, was the Thursday killing, just the week before last, of five men and three women in the shambles and bloodletting around the Bastille quarter between the gendarmerie and about twenty thousand anti-O.A.S. marchers, largely Communists, whose march had been forbidden by the government—some of the dead dying of suffocation or from internal injuries inflicted by the pressure of the crowds when they panicked before the gendarmes' clubs, one or two dying of head wounds, and the youngest, a boy of sixteen, being trampled to death. It was an evening of shameful and historic savagery in the capital of France, comparable to the night of February 6, 1934, when on the Place de la Concorde the Third Republic, faced with a collapse of the government and threatened by both Right and Left, ordered its soldiers to fire, killing eighteen in the dense, excited crowd around the statues and fountains.

On the Tuesday morning following the Thursday deaths came the gigantic funeral march for them, of a million marchers, or a half million, or a quarter million—nobody seems to know exactly how many, nor could we who were there count ourselves, knowing only that on the Boulevard du Temple we packed the streets and sidewalks solid all the way from the Place de la République, which was also overfilled, to the black-draped Bourse du Travail, where the coffins lay, and which was turned into a *chapelle ardente*. The street was almost gay with men carrying official *couronnes mortuaires* and set pieces of flowers bearing red ribbons with the names of the donors in gold, such as the Railwaymen's Union of Vitry, a Paris suburb, or a local printers' syndicate. It had rained during the night, and the air was moist, making the flowers glisten freshly in the big floral constructions—flower beds of lilacs, arum lilies, and scarlet carnations, which two men usually carried between them. There were little humble bouquets in the hands of the women and girls, mostly of multicolored anemones and yellow mimosa, still fluffy in the chill atmosphere. Other girls were selling small black-bordered cards bearing the words *"En Hommage aux Victimes du 8 Février, 1962,"* which marchers pinned to their coats, paying whatever they chose, the money going to the victims' families. Seven of the eight dead, including the three women, had been Communist Party members. The husband of one, Mme. Fanny Dewerpe, had been killed in the Communist riot organized against the 1952 Paris visit of General Matthew Ridgway, then being accused by the Left of having exploded bacterial bombs in Korea. In the European processional manner, an enlarged photograph of each of the dead was carried on a pole by a member of the family, who headed the procession as it finally started for Père-Lachaise Cemetery (which those marchers at the end of the procession reportedly did not reach until four rainy hours later). The photograph of the adolescent, Daniel Féry, who had been a handsome lad, was carried apart and alone by a boy who had been his best friend. Daniel was an apprentice of some sort working for *Humanité,* the Party paper. His father told one of its reporters, "Whenever there was a street demonstration, he always attended. He liked to exteriorize [*sic*] himself. I suppose I should have been firmer with him. I would have been if I had known it was going to kill him."

By the end of 1961, six hundred plastic-bomb explosions all over France, many of them here, had damaged a lot of property. A list of three thousand addresses to be plastiquéd in 1962 was recently found on an O.A.S. agent when arrested. So far, there is no official count of the January and February thousands of bombings throughout the country, or of the hundreds in Paris. The big French question now is: Who is going to pay for the damage? French fire-insurance policies, which are apparently what bombs normally come under, do not cover acts of civil war, which are apparently what the O.A.S. explosions come under. (The insurance companies have already prepared a clause refusing responsibility for atomic bombs, too.) By now, there have been so many explosions in Paris that one is bound to know people or apartment houses or shops that have had their bomb. The destruction is freakish, and can be utterly shattering. In the elderly apartment house that was bombed early in January because Jean-Paul Sartre had a flat there, the staircase and stair well were blown to smithereens halfway up, the whole central open spine of the old building was damaged, and doors were splintered and walls cracked. Repairs are estimated at twenty thousand New Francs. It is coöperattively owned, and probably Sartre is the sole tenant who is well enough off to pay his share of the damage. (He has just lodged a complaint at the Palais de Justice against "X" for an attempt on his life.) For the other literary tenants, such as Mme. Béatrix Beck, a former Goncourt Prize winner but no best-seller, such financial participation is out of the question. She and the other upper-floor tenants now go up and down by the service staircase. Just behind your correspondent's hotel, last week at dusk, a bomb went off in the Rue du Mont-Thabor on an apartment building's inner-court staircase—always the favorite spot, with a greater chance of material destruction but a lesser chance of costing life. Every window in the house was shattered. The blast, illogically, also blew out the glass street front of a dry-cleaning shop, leaving the clients' garments hanging in perfect order on their racks; the young concierge was cut about the throat, and her baby injured. Anyone walking by on the sidewalk in front of the cleaner's would have been pierced like a colander. The shattered glass was ankle-deep all over the street. By the next noon, a glazier had most of the new court windows in place. Glazing has become an active revived profession here. Some journalists especially distinguished for their anti-O.A.S. sentiments have been bombed twice over, with private libraries or bookshelves ruined, the books' spines blown off, the glue blown loose, and volumes sometimes blown clean out of the windows.

The carelessness and inefficiency of the O.A.S. bombers in some cases have added to the angry contempt in which they are held. They mix up addresses. Recently, they bombed an apartment house at No. 13 on a certain avenue near the Invalides when they meant to bomb No. 15, where two Gaullist deputies lived. Nice young married couples used to try to rent flats in apartment houses where a deputy lived, since his presence gave tone, and even security, to the whole establishment; now mothers and mothers-in-law warn them to stay out of such bomb traps. M. Sartre, who had early received menacing letters from the O.A.S., had actually left his flat two months before it was bombed, and, on top of that, the bombers got the wrong floor when they planted their plastic. The other day, the O.A.S. bombed the apartment of a Communist deputy who has been dead for two years.

March 8

In the course of the cease-fire talks that are finally going on at Evian-les-Bains, the leaders of the Algerian Nationalist Party have unexpectedly asked if the French government can control the murderous unofficial violence of the O.A.S. against the Algerian population, the O.A.S. now being where the gunfire comes from—usually casually, on the main streets of the cities. This question turns the original French thesis for peace upside down, since that thesis was based on the premise that the French whites would have to be protected from the liberated Algerians. In Monday's pre-dawn shower of a hundred and seventeen O.A.S. plastic bombs in Algiers, twenty-seven of the thirty-five people killed were native Algerians, such as, over the months, have supplied from three-fourths to four-fifths of the daily dead. If the embittered Moslems, overprovoked by the O.A.S.'s small, unrelenting day-and-night massacres, burst from the Casbah with their knives drawn, the peace will be lost before the dove ever becomes the bird in the hand.

At the moment, hope still centers on Evian. In the weekly *Express,* the mordant, phantasmagorical artist who signs his cartoons Tim lately had a picture illustrating the cease-fire, after seven and a half years of war. It showed a one-legged, one-armed French soldier leaning on a crutch and holding a bugle to the mouth of a bugler who has both legs but no arms.

March 23

The curé of a small village near the Channel coast is reported to have rung his church bells at dusk on Sunday to announce the good news of the cease-fire in Algeria to his little countryside—apparently the only tintinnabulation of the sort noted anywhere in France. Here in Paris, the expected seven-o'clock Sunday-evening official radio announcement of the truce in the Algerian war seemed to give people a sense of profound relief but elicited no signs of animation, as if it had been awaited so long—and, lately, so often—that it was already too familiar as a hope to need the panoply of sudden flags on balconies or cheering street crowds, of which there were none, or the sound of the great Bourdon bell in the belfry of Notre-Dame, also silent. Obviously, the emotional satisfaction at the truce was troubled even in advance by doubts about how it can or will work at first. Also, this satisfaction was greatly debilitated by the fact that, with enormous difficulty and endless litigious arguments, the truce finally bound over only the two original belligerents in the war, the French and the Algerians, whereas behind them, for more than a year now, has been the constantly increasing third force, the uncontrolled enemy of them both and of the future peace as well—the random, deadly, implacable power of ex-General Salan's Organisation de l'Armée Secrète, at least as it operates in Algeria. President de Gaulle, in his television speech to the nation at eight o'clock Sunday night, referred with supreme and angry disdain to certain former Army elements as "misguided chiefs and criminal adventurers" (with a stately kind of linguistic purity, he never permits the name of the O.A.S. to cross his lips), but his millions of listeners all over the country knew only too clearly whom he meant, and may well have wished that he had gone on to say what he planned to do about them at this eleventh pacific hour, and about those who favor them. *Le Monde,* in its front-page Monday-night editorial, titled "Au-delà de la Guerre" and signed "Sirius," nom de plume of its editor, Hubert Beuve-Méry, who takes to his pen only on dangerous or vital occasions, stated drastically, "What above all must finally cease are the evident complicities [these men enjoy]. When these 'misguided chiefs and criminal adventurers' make themselves masters of the streets and of the commissariat with ease [as they have done in Algiers and Oran], receiving food, supplies, and arms by train and by truck, there is ground for declaring that they benefit from alarming collusion and, without doubt, from powerful aid. If this does not change, the worst fears are justified. Citizens of a weak state that is more and more powerless, the French will indeed find themselves plunged into a civil war."

Half an hour after de Gaulle spoke on the air, Premier Benyoussef Ben Khedda, of the Gouvernement Provisoire de la République Algérienne, spoke from Tunis to the Algerians in Arabic to announce the successful conclusion of the Evian conference ("It is a great victory for Algeria"), and to order the cease-fire. In closing his speech, he bluntly warned, "The period of transition will demand the greatest vigilance. The cease-fire is not the peace. The danger is great. The Fascist hordes and the racist-minded O.A.S., despairing of maintaining their French Algeria, will try once more to cover the land with blood. Up to now, the French civil and military authorities have been more or less accomplices of the O.A.S. In the superior interests of peace and of coöperation between the two countries, this complicity must come to an end." This final item of official bold comment and advice from the newly former enemy was omitted by certain Paris newspapers—even by papers that had themselves earlier expressed the same worried criticism of the de Gaulle government, if more tactfully.

On Monday, each of the papers gave a full page to the detailed and lengthy contents of the Franco-F.L.N. Evian accord, in which the elaborate chapters concerning the statutory position of Europeans in Algeria were, of course, of major interest to French readers. In Paragraph 1, titled "Dispositions Communes à Tous les Algériens"—which includes white Algerians of European stock—the opening clause, on "Personal Security," declared, "No one shall be disturbed, hunted, pursued, or condemned, or made the object of penal judgment, of disciplinary punishment, or of any discrimination whatsoever because of acts committed in relation to events that occurred in Algeria between November 1, 1954 [when the war began], and the day of the cease-fire." The second clause read, "None may be disturbed, hunted, pursued . . . because of words or opinions uttered in relation to Algerian events between November 1, 1954, and the day when the Algerians will vote on their self-determination." It is thought that this vote will be in July, until which time the Algerian settlers can say anything reckless or vilifying that comes into their heads—for instance, about General de Gaulle, whom they hated before for having promised to end the war, and now hate boundlessly for having patiently preserved his word of honor. In clarification, the next paragraph, entitled "Protection of Persons and Goods," states that "French nationals"—i.e., the white settlers, or *pieds-noirs* of European stock—who "exercise Algerian civil rights" as residents "will benefit by the same measures and under the same conditions as in Paragraph 1." This would seem to mean that there is to be an amnesty (though that word is nowhere used in the text) for the raging, murderous participation in the O.A.S. terrorist campaign of formerly ordinary, normal adult white citizens who have run amuck under the recent pressure of events and become lawless, bloodthirsty monomaniacs in their hope of keeping their Algeria French forever, and for the blood-drunk *ratonades,* or rat hunts, on the Algerian city sidewalks, in which rich bourgeois *colons'* sons shot down passing *ratons* (the old settlers' name for the poorer Algerians) in the sadism of youthful bravado—an amnesty, in short, for all these human degradations and horrors, commonplace or spectacular, and many dating from only last week. The realistic, historical wisdom behind this implicit amnesty—or so some French feel—is that there is no difference between soldiers and civilians in a bitter civil war; both kill.

However, the French government's first act since the cease-fire has been to create courts-martial in Oran and Algiers, empowered to judge, by the most rapid procedure, any adult French citizens who murder or throw bombs and are caught in the act. So far, no one seems to have been caught *in flagrante delicto,* doubtless owing to the extremities of confusion, hope, bloodshed, and violence and the unaccustomed attempt to set up order down there on two levels—that of the Moslems, whose disciplined conduct is now officially in the hands only of the Algerian F.L.N. authorities, and that of the French and European whites, whose conduct so far constitutes the bigger danger to peace. By yesterday morning, about fifty dead and nearly two hundred wounded—mostly Moslems, as usual, and all struck down since Monday noon's declared cease-fire—had already been reported here in Paris from still anguished Algeria, in its first bloody peace throes.

A multitude of secret documents detailing ex-General Salan's plan of action after the cease-fire were seized only last week—just in the nick of time—at the Résidence de St.-Raphaël, in El Biar, in the hills above Algiers. In order to have these papers boomerang against the surprise on which the O.A.S. was clearly counting, the French authorities have intelligently broadcast them in Algeria, in Arabic and in French, and have given them out for publication on both sides of the Mediterranean.

His secret papers reportedly said, "I want a general offensive, an extension of the revolutionary war that we are waging. The aim is not to attempt a definite *Putsch*." What Salan did want was to multiply the O.A.S. subversive action everywhere, and "by favor of this chaos" to replace little by little the legal authorities with O.A.S. power, and he added, "In brief, we must rot conditions to the maximum, so that Algeria will at the end fall into our hands like a ripe fruit." As for his precise elements of action, he said that the O.A.S. should play on the city dwellers' emotional reflexes. One of the first objectives would be "to asphyxiate, to stifle, the urban centers by strikes." The *bled,* or countryside, he considered unimportant; it has not shown itself to be "valuable ground for revolutionary action." He added, "A *maquis* does not do much good." His dominant principle, as revealed in his secret papers, was to develop the subversive war in every way, to institute massive shutdowns of all kinds of work and other normal activities. On the purely tactical side, he said that his commandos had to multiply "their harassing of the forces of law and order, so as to wear down their nerves—to weaken them, if breaking them proves impossible." Also, the commandos were to fire on the gendarmes and the C.R.S. (the French shock police), burn their vehicles and gasoline stations, and pour oil on the streets, so the C.R.S. trucks and tanks would skid; "that way, our men can more easily attack them." At its end, the Salan plan got around to the native Algerians as a final human domain for the O.A.S. psychology. According to *Figaro's* transcript of the Salan secret papers, "All the Moslem cadres must be attacked with increasing vigor—doctors, lawyers, druggists, civil servants, technicians, and so forth. The object is to destroy the best Moslem elements in the liberal professions, so as to oblige the Moslem population to turn to us."

On Thursday afternoon in Algiers, the O.A.S. pasted on walls warnings addressed to the French C.R.S., the gendarmes, and the French Army soldiers on patrol duty, saying, "You have until midnight tonight, Thursday, March 22nd, to withdraw from the Bab-el-Oued quarter. After that, you will be considered troops serving a foreign country. The cease-fire of Monsieur de Gaulle is not that of the O.A.S. For us the fight begins." It actually began a few hours earlier, in the first all-out O.A.S. attack, with heavy weapons, on Algiers and its Bab-el-Oued district. Apparently, the civil war down there has truly started.

On Tuesday, the French Parliament opened for three days, in a tense atmosphere, to listen to President de Gaulle's special message to them, read standing by the Chamber speaker and heard standing by all the deputies except the Socialists and the Communists, who remained seated on their benches as a sign of disrespect. No vote on the message was allowed. Though this Chamber is a remarkably biddable one, de Gaulle will doubtless soon dissolve it, and call for new elections after his April 8th referendum, on which he should win a ninety-per-cent "Yes" vote. The opening speech in response to the message was by Pierre Portolano, deputy from Bône, in Algeria, and chief of the Unité de la République political group, which is pro-Algérie Française but anti-O.A.S. and anti-de Gaulle, and some hundred deputies rose to hear him—an impressive array. His was an intelligent, unhysterical, strong speech by an Algeria-born right-wing diehard. He deplored the handing over of an undeveloped Algeria, to become, in its independence, a pro-Communist popular republic (he feared), and he deplored the lost rights of Parliament, "baffled by General de Gaulle." To Premier Debré, on the Government bench, he delivered a telling shot by saying, "Your former words still remain in our spirit: 'The abandonment of Algeria is an illegitimate act' "—which, of course, Debré, who had early been pro-Algérie Française himself, had indeed said. At the end of the morning session, the one-time Poujadist deputy Jean-Marie Le Pen, a former paratroop officer and still a hotheaded bully, said to the Minister of Overseas Departments and Territories, Louis Jacquinot, a former lawyer, "You'll all be hanged," to which His Excellency replied, "You might need me someday to save your own head." In the night session, Le Pen accused the absent de Gaulle of using Algeria to "strangle the Republic and set up a dictatorial empire."

"The mere fact, M. Le Pen, that you can speak so intolerably proves the liberties of this government," the Chamber speaker replied.

They were rather sad farewell sessions.

Parisians, familiar with the unabating, witty, insulting anti-Gaullism of France's only fearless Leftist satiric weekly, *Le Canard Enchaîné,* were astounded to see its sturdy *amende honorable* in the present number, headed boldly, "To de Gaulle, from His Grateful Country: Once and for All, MERCI!" The impertinent, intelligent little editorial that followed said, in part, "It is extraordinary how many times we have had to shout *'Vive de Gaulle!'* or *'A bas de Gaulle!'* We admit it, it's crazy. Yet when a gentleman says one day to his nation, 'Algeria is a French land today and forever' (June 6,

1958), and another day proclaims 'Algeria will be a sovereign independent state' (October 2, 1961), that is rather crazy, too. Well, let's pass over all that. To the man who would bring us on a silver platter the end of the Algerian war, whoever that man was, we were ready to take off our hat. You could say that we've been waiting for that man. It could have been Guy Mollet, but it wasn't. It could have been de Gaulle in 1958, de Gaulle in 1959, de Gaulle in 1960, de Gaulle in 1961. But it wasn't any of those de Gaulles. It could only have been Salan in 1964, if he had staged a successful insurrection in 1961. However, at the end of it all, it was de Gaulle just the same. *Alors, 'Vive de Gaulle!'*" With the same grateful spirit, if not the same gaiety, millions of French citizens feel the same way: *Vive de Gaulle!*

April 3

Paris will now go back to politics, or as close to politics as the French ever get under President de Gaulle. The French are invited to vote next Sunday in a referendum, which will be like voting under a blanket. In his own recent words on television, de Gaulle asked them to give him "a massive affirmative response," declaring their adhesion to him "in my capacity as chief of state"—a strange but not unaccustomed request on his part for direct homage to that double identity in his eyes, France and himself.

April 18

According to the latest political French joke from London (no such quips are native to Paris these days), President de Gaulle fixed himself up with such extraordinary added powers in his last referendum that if he wanted to he could change a man into a woman. Actually, the only thing he has changed is his government. Over the weekend, it was changed for the first time since he came to power, in 1958. It was so long-lived a government that it lasted three years and ninety-eight days—the most durable republican government known to French history. With it went Premier Michel Debré, its head, also a historical record-breaker in that he was the most unpopular Premier of modern times. Speaker Jacques Chaban-

Delmas of the Assemblée Nationale compared Debré to St. Sebastian, martyred by the arrows of fate, which were continually being shot by his enemies into his small, pugnosed person, many of them symbolically aimed at President de Gaulle himself, against whom few have dared really draw the bow, Debré being a more vulnerable and convenient target. His disappearance and that of his veteran government, after these three constructive, vital early Fifth Republic years, took place quietly within a mere half hour, just before noon on Monday, in the Hôtel Matignon, where Debré shook hands with, and gave way to, the equally quiet incoming Premier, Georges Pompidou, at whom the extreme French Left, consisting of the Communists and the Socialists, have already dutifully started taking pot shots because he is director-general of the Rothschild bank. In this rapid, civilized change of governments, there were none of the customary preliminary offstage political noises—the crash of the falling regime or the panting of the newcomers scrambling toward brief Parliamentary leadership—and no sign of the regulation chaotic, anarchistic rumpus devoted to the carrying on of republicanism which the French people became so familiar with in the cynical, elderly Third Republic following the First World War, and which the younger French had to learn by heart as their main political lesson in the ill-starred Fourth Republic after the silence of the Nazi Occupation. Although this new Pompidou regime is only the second government that the Fifth Republic has installed, it is the twenty-seventh since the Liberation, which gives one an idea of the turmoil and turnover that de Gaulle's high and autocratic governing hand has spared his *belle France*.

The new government is, by chance, the most literary that modern France has known—another odd contemporary record. Son of a schoolteacher in the Cantal region, which is known for its delicious large, pale, solid cheese, M. Pompidou was educated at the Ecole Normale Supérieure, where the élite of the French teaching profession is recruited; took his degree in literature; was a professor in his subject in Marseille and also at the famed Paris Lycée Henry IV; has published studies on Britannicus, Taine, and the works of Malraux; and recently brought out a pleasant anthology of poetry, Baudelaire being his favorite poet. He also gave some respectful editorial advice to de Gaulle in the compiling of his official memoirs. Pompidou has been an associate and a devoted counsellor of de Gaulle's, off and on, ever since 1944, when the General became head of the Provisional Government, part of his constant value being that he has never been in politics, as either deputy or senator. He was secretly used early last year in undercover meetings in Switzerland with the rebel F.L.N. leaders, and prepared the basis for the truce negotiations, for which others got the credit. In all these various activities, de Gaulle has, in complimentary fashion, referred to Pompidou as "my signature," and he is already being called "the master's voice." With this non-political Premier, the Fifth Republic's executive power will be centered in Parliament even less than it was under Debré and more than ever in the Elysée Palace. Pompidou's apolitical career may make him more sympathetic than was Debré (a former senator and thus legislative-minded) to de Gaulle's announced faith in referendums as "the most frank and democratic method" for the ordinary citizen to use in discharging his political responsibilities—or so de Gaulle told his citizenry on television just before he held his last referendum, ten days ago. This was a statement which to the intensely partisan-minded French politicians was sheer historical heresy on the General's part. They would have fought against it in the new elections that it was thought de Gaulle would call this spring but that he has, with far-sighted statesmanship, postponed to the spring of next year.

Maurice Schumann, who was *porte-parole* for de Gaulle's Free French on the British radio during the war and is now the new Minister for Regional Planning, has recently written his first novel—"Le Rendez-vous Avec Quel-qu'un," which purports to be the confessions of a German S.S. man—so he is second on the new Ministerial literary list. The writings of Pierre Pflimlin, the Minister for a new post called Coöperation with the African States, have been more technical, since he comes from the northern textile city of Roubaix; one of his books is "L'Industrie Textile Alsacienne." Alain Peyrefitte, the new Secretary of State for Information, dropped his first name of Roger so as not to be confused with his cousin Roger Peyrefitte, who is the author of "Les Ambassades" and other semi-scandalous books occasionally put on the Vatican Index; Alain Peyrefitte has himself written much less talked-about novels.

Many of the ten new Ministers are professional politicians, and some were early professional Gaullists, like Schumann and like Gaston Palewski, the new Minister for Scientific Research and Atomic and Space Affairs, and until lately the French Ambassador to Italy, who was an admirer of Captain de Gaulle and his new mechanized-army theory (which later meant tanks) as early as 1931, and who, in June, 1940, was one of the first to join General de Gaulle in London. These are men whose loyalty and lengthy political experience could go a long way toward making this new government a realistic body, to offset the Elysian relations in the Palace between today's President de Gaulle and Rothschild's Pompidou. The truce in Algeria marked the end of the principal effort to terminate the Algerian war, which brought de Gaulle's Fifth Republic into being and de Gaulle himself into his phenomenal power. His new government marks a second stage in his republic, and a new stage in the ripeness and in the coming struggles of the President himself.

May 1

When President de Gaulle's new Premier, M. Georges Pompidou, well-dressed in blue, with what looked like a white silk summer tie, presented himself last Thursday morning before the National Assembly, he was a French public figure without historical precedent. He was the first in France's five republics to be merely a civilian personality who had never belonged either to the Assembly or to the Senate and had now come before the deputies as head of a government. Architecturally as well as politically unfamiliar with the scene, he hesitated, as if unsure of his proper place, before seating himself on the government bench in the first row, that seat of power which so many backbenchers know so well from having vainly stared at it for years with envy. When he started reading his speech from atop the speaker's tribune, his voice sounded as if he felt moved at hearing it there; then, with steadiness, he began enumerating the new government's program of de Gaulle ideas, which, as a long-time confidant of the President's, he must have known by heart. Though his Premiership was a de Gaulle personal appointment, he tactfully requested of the deputies a vote of confidence, which, in a manner of speaking, he won the next day by 259 votes to 128, qualified by 119 abstentions—a result that many deputies interpreted as 247 votes against de Gaulle and only a few more than that in favor of his Premier. By that time, M. Pompidou had undergone a lengthy, heckling question period, lasting until late Thursday night, at the hands of thirty deputies who were dissatisfied with de Gaulle's policies, and had also endured two of those dramatic Gallic interpellations—part wit and part political anger—that often furnish Parliamentary *coups de théâtre.* One anti-Gaullist deputy of the extreme right, denouncing the "progressive abasement of Parliament," suddenly said to the new Premier, "Your predecessor, Debré, was an easy target to shoot at. You are less vulnerable. You have not expressed your opinions on anything, if it so happens that you have any. You are a new man. Profit by this advantage; it will not last. Virginity, like matches, can serve only once." Then came the most vigorous Chamber critique of de Gaulle's government of "benign personal power" yet heard in the Fifth Republic, made by one of the Chamber's oldest republican practitioners—eighty-three-year-old former Premier Paul Reynaud, still a dominant, tiny figure, hardly taller than a large dwarf, for over thirty years de Gaulle's most durable Parliamentary friend, and until Thursday still thought a staunch Gaullist. (It is rumored that he had loyally warned the General in advance that their political friendship was now over.) To Pompidou, Reynaud said, "The proof that we are not in a Parliamentary regime is that you are here, Monsieur le Premier Ministre. You have already been thrown from power under Louis-Philippe; at that time, you were called Count Molé." At the aptness of this Restoration fantasy, the deputies began to laugh. "He was a very distinguished man, like you. Completely unknown to the general public, like you. One whose political opinions nobody knew, like you. And who had the entire confidence of the King, like you." At this reference to de Gaulle as a monarch, the Chamber, except for the discomfited Gaullist U.N.R. Party, burst into applause and louder laughter. "His majority was so derisory that he could keep power only by dissolving the Chamber. But on facing the people's vote he was defeated, despite indecent official pressure"—this last being an angry reference to the new Premier's having been personally appointed, instead of selected in Parliament by politicians representing the people. Then Reynaud railed against the Fifth Republic practice of taking referendums, leading merely to a Yes or No, which short-circuits endless Parliamentary debate on vital matters, and he added, "Is not the proper essence of a Parliamentary regime the formation of statesmen? Take care! When the Constitution is not respected, there is no longer a Republic." At the end, he received a salvo of applause.

Reynaud had begun by asking, "What will happen when de Gaulle is no longer here?" What a majority, perhaps two-thirds, of the ordinary French people seem to fear is that there will be a Sixth Republic identical with the Third and Fourth Republics, in which governments will rise from and fall back into Parliament's hemicycle like boomerangs, and power will again be purely political and splintered. (There are seven political parties in the somnolent Chamber right now.) It will no longer be personal, benign, semi-autocratic, and ultra-patriotic—qualities that, with a new government in focus, anti-Gaullist sentiment has now openly spoken against.

After the sensational arrest in Algiers on Good Friday of ex-General Raoul Salan, the police photos of him rigged up for clandestinity with black hair and a cavalryman's mustache were a considerable shock to the Parisian public. They had had an image of him, in his infamy. In these police pictures, not only did he look unlike the way they had seen him pictured for an exact year, lacking five days, as the Algerian O.A.S. chief and the supreme enemy leader against the French state—smooth-shaven, with white hair above his odd, unsymmetrical eyes and inscrutable expression, like an elderly, pessimistic silver fox—but he did not even look alive. He looked as if the death sentence that he will certainly be given at his coming trial had already been carried out and here was the proof of it—the posthumous-looking, inanimate wax dummy of himself, readied for the Musée Grévin's chamber of horrors. The other immediate public reaction to his arrest was that it would probably save his subaltern, ex-General Jouhaud, recently condemned to the firing squad. At first, perhaps because both were housed in the Santé Prison here, there arose a feeling that the two of them were irresistibly united, and that both should be shot for all the lives and limbs they had ordered destroyed. Now, more reasonably, has come a realization of the appalling, typically French paradox and the dreadful delicacy of de Gaulle's problem in his coming decision on Salan. It was Salan's insurrectional cry against the Fourth Republic of "Vive de Gaulle!" four years ago, on May 13th in Algiers, that brought de Gaulle to supreme power as head of France today and gave him the right to reprieve Salan that he will use or fail to use before the end of this month.

The trial should open in three weeks and last for a week. It will be the most important and disturbing trial in modern France since that of Marshal Pétain, because once more it will involve the Army, which, though waning and much declined, is still the greatest emotional cult in France except for the Church. Because Pétain disdained his trial as illegal, he sat through it without speaking, as if he were not there. Salan may be silent at his trial, too, as he has already been in the presence of his *juge d'instruction,* who on Monday completed a last series of monologues with him, representing the more than two hundred questions he was required by law to ask in preparation for the trial. "If I do speak in court," he has let it be known, "it will not be for my own sake. My life is finished. It will be for the sake of history." His secret code name as O.A.S. chief was Soleil, the Sun, carrying with it an undertone reflection, like the false shimmer in Salan's *folie de grandeur,* of Louis XIV, the Roi Soleil.

The psychological portrait of Salan now current in Paris is elliptical but fascinating. A complex personality, he was considered throughout his career and until recently only a brilliant second-in-command. Protestant in religion, son of a poor tax collector, and brought up in Nîmes, he was a Meridional without warmth, devoid of magnetism but endowed with amazing sangfroid and with a lucidity sharpened by peril whenever at a dangerous turning, so that, when younger, he always pulled the right string to unravel a situation, political or military. In brain, style, and *panache,* he never held a candle to his long-time chief in Indo-China, the aristocratic little martinet General de Lattre de Tassigny, to whose command there he eventually succeeded. In the East, he was nicknamed Le Mandarin, partly because he was secretive and enigmatic, partly because, like many French officers there, he reportedly smoked opium. He was described as melancholic, somewhat majestic, using his hands in slow gestures, one finger ornamented with an elaborate ring. Most of his career was in the Far East—Tonkin, North Vietnam—with an earlier stretch in North Africa; in any case, he was always far from France, in the dust of outposts. He negotiated with the Vietnamese and with Chiang Kai-shek, held conversations with Ho Chi Minh, was considered an ace in the Deuxième Bureau secret service, was brave in combat, and became the most decorated officer in the French Army, with fifty-seven decorations indicated on the board of ribbons he wore on his breast. Apparently, the great alteration in his military and political vision came in 1954, when he was sent back to Indo-China to investigate the circumstances of the French catastrophe in the fall of Dienbienphu. There he became a prey to the *complexe de l'abandon*—the embittering belief that the modern French state invariably deserted its Army, thus forcing it into defeat and the empire of France into another humiliating loss. This morbid, romantic military idea spread like a contagion among the French officers and was basic to the organization of the Secret Army in Algeria, where the civilian white French population was already suffering from similar delirious symptoms. Shortly after this, the blood began to flow in the streets of Algiers and Oran, and the deadly plague of the O.A.S. civil war was started. Salan's arrest has come dangerously late in the Franco-Algerian peace plans.

May 17

The prisoner spoke only once on Tuesday afternoon at the opening of his trial in the Palais de Justice, when, as is customary in a French court, he gave his full identity and particulars at the judge's request: "Salan, Raoul Albin Louis, born June 10, 1899, in Roquecourbe, ex-general"—for the disqualifying prefix he raised his low voice to make it distinct—"of the Colonial Army, Grand Cross of the Legion of Honor, Military Medal, Cross of the Liberation, wounded in action." It was the only time he was to speak during the five hours of the proceedings, though his pessimistic mouth kept moving almost continuously, his thin lips active in silent nervousness. His pallid face had an emptied, impressive look of lost and dissipated energy, odd beneath his clownlike hair, grown freshly gray over his ears but still a mawkish henna color on top—a leftover from the dye he had been relying on as disguise when arrested in Algiers. Gradually, he began covertly looking around, reconnoitring, in a slanting glance taking in, to his right, the imposing High Military Tribunal, with the president of the court, the prosecutor, and the court clerk in their pseudo-ermine capes and scarlet robes; then he looked more vaguely at the French press box opposite, where the journalists sat staring at him and taking him down as a portrait in words. He appeared not to listen when the clerk rose and read aloud the pair of long reports to the President of the Republic—the special form in which the charges against him were contained. Like two connected scenarios of Salan's recent subversive private military life, which brought him to where he sat, inattentive and overfamiliar with their story, the first dealt with the action of the unsuccessful 1961 spring *Putsch*, and the second, like a sequel, with the ensuing

O.A.S.—the elaborate, irrational, and dangerous product of his imagination, sectarian patriotism, Army training, and egotism. Rapidly read aloud as a series of facts in an enormous, overcrowded courtroom, and listened to while their creator sat in the prisoner's box, smartly enough dressed in civilian slate gray with a diagonally striped black-and-blue tie, his O.A.S. projects, including their indubitable successes to date, seemed like nothing but incredibly well-planned, sanguinary lunacy. Salan was accused of leading an armed revolutionary force against the Republic's institutions and of inciting citizens to take arms against each other, which constitutes civil war, and for which the penalty is death.

Salan's power of disturbance continues. On the *quai* leading to the Boulevard du Palais and before the wings of the Palais itself, iron barricades have been set up to keep people and possible plastic bombers off the sidewalk. Military police with small machine guns are stationed every few feet, and all but one of the high, ornate grilled gates to the Palais outdoor courtyard have been closed. At this gate, identity papers must be shown to armed guards (even by outraged lawyers bent on private business); inside the courtyard, full of gendarmes and police cars, the Sainte Chapelle, that jewel of medieval glass, is closed to visitors throughout the trial (as are five minor Palais courtrooms near the trial room). Five hundred journalists from all over France and the Western world requested tickets for a courtroom that has press space for perhaps an uncomfortable two hundred. In a judgment of Solomon, the Presse Judiciaire bureau gave the newspaper people all the hard benches normally claimed by distinguished visitors. Inside the Palais, one must show a court ticket and a professional press card to military-police officers, who courteously ask permission to search lady journalists' handbags; the men are equally politely frisked. To avoid possible danger in driving Salan to and from the suburban Fresnes Prison every day, he was transferred to the Palais on Wednesday at dawn, to be lodged in a basement section for temporary prisoners, where he has been inexplicably put in the women's wing, under the care of the Sisters of St. Vincent de Paul. His is the small room in which Pierre Laval vainly tried to commit suicide after having been given the death sentence in the famous assize court above, where Salan is being tried—where Marshal Pétain was tried.

This being the most important political trial in France precisely since that of Pétain, it has been well understood in legal and political circles that Salan's lawyers could try to defend his life only by prosecuting the Fifth Republic, for which his team of three lawyers was perfectly chosen, all being rabidly pro-Algérie Française. It is the first historically important case to fall into the gesticulating, theatrical hands of the trio's chief, Maître Jean-Louis Tixier-Vignancour, wellborn, highly intelligent, with a Mephistophelean basso voice and a semi-infernal wit. At once on Tuesday, he began his delaying and destructive tactics by impugning the judicial competence of this High Military Tribunal, among other things complaining of an insufficient preparation of the prisoner by the *juge d'instruction.* He declared that Salan had been given only "three days, reduced from the three months accorded Marshal Ney"—Napoleon's "bravest of the brave"—"who was shot." He added, with an insolent glance at the bench, "But a Paris boulevard was named after him. The magistrate who handled the affair did not even have his name put on an alley."

Yesterday, Tixier-Vignancour demanded the presentation in court of more than forty of the witnesses he had already cited, as another annoying tactic of delay, and, with unexpected stage-managing, he had them brought into court by a back door and introduced to the judge by a *huissier,* who asked their names and repeated them to the judge like a butler at a fashionable reception, after which they were hustled off into a waiting room. It was a disruptive and astonishing scene. Among the witnesses were the still beautiful widowed Mme. la Maréchale de Lattre de Tassigny, dressed in her permanent, elegant melancholy black and a chic black hat; a parade of generals and colonels, some of whom bowed cordially to the prisoner on the way out; high civil servants; a former Fourth Republic Minister; some extreme-right-wing deputies; one witness whose name nobody in the press section caught and who was manacled to a police guard; and, suddenly, a debonair young Army captain, wearing his cap and immaculate white gloves, who, on being introduced, smartly saluted the judge (who bowed, dazed), pivoted on his heel, and equally smartly saluted Salan (who, for once, smiled, and could afford it), and then removed his cap and gave the judge his news. He declared that, newly on duty in Lille, he had been telephoned long distance by a commanding officer five minutes before he started for court and forbidden to appear as a witness. But since his subpoena had declared that his failure to appear was punishable by law, there he stood—*"et me voilà!"* This example, following that of an admiral and a reserve general who also declared that they had been warned by phone by the Chef de Cabinet du Ministre des Armées not to answer their summonses as witnesses, adding that the official had told them the court president concurred in this (which, in some confusion, the president denied)—all this gave Tixier-Vignancour his tremendous chance. His huge voice lifted in a cry of indignation, and he launched into a diatribe, declaring that the Chef de Cabinet must be called as a witness, too, and that within the hour a charge would be made against Pierre Messmer, Minister for the Armed Forces, accusing him of suppressing legal communications and of breaching Article No. 173 of the penal code. While this was going on, crescendo, Tixier-Vignancour received word, and passed it on to the bench, that the young captain in the waiting room had just been ordered by an Army messenger to report at once to the office of the Chief of Staff of the Armed Forces. All this produced an exceedingly unpleasant impression, especially on the many lawyers in their robes who had dropped in to listen to this important trial. It was felt that the Ministry of the Armed Forces had badly served the standing of the Fifth Republic, and might even have helped save Salan's head, were that possible, which no one believes it is.

It was deduced that the reason Tixier-Vignancour had dawdled on Tuesday afternoon, leaving Salan no time to make his personal declaration—the crux here of any great trial—was so as not to have his prisoner diminished by sharing the front pages of the French and world press with President de Gaulle, who, to the surprise of many of the French people, had chosen to dominate the news that particular afternoon by giving, in an Elysée press conference, his own personal declaration on Europe, the Common Market, German relations, President Kennedy's United States, and, of course, de Gaulle's France. So it was not until late Wednesday that Salan was heard from. In a dry voice, and wearing gold spectacles, he read for almost an hour a report on himself that certainly he never wrote but that, in its few personal statements, which he must have dictated with bitter emotions and memories, was impressive. *"Quand on a connu la France du courage, on n'accepte jamais la France de l'abandon,"* he declared, in part. "From the beginning, self-determination for Algeria was merely a lie destined to cover abandonment of that country." He said he was convinced that on May 13, 1958, when the rightist officers in Algeria first rebelled against the Paris government, he had been "the dupe of a frightful sacrilegious farce." Yet on May 15th, "I chose to bring de Gaulle back to power. If I misled the people of Algeria and the French Army, it was because I myself was misled. It is [today's] government that, denying its own origins, is responsible for the blood now flowing, and more responsible than anybody is the one I gave power to"—meaning, of course, de Gaulle. "I do not have to exonerate myself for not wanting to see Communism install itself an hour away from Marseille, and Paris within distance of its short-range rockets. I do not have to exonerate myself for having defended the wealth that young pioneers have given France in the Sahara, assuring its independence in oil. If the Allies had lost the war, the Germans . . . would have loudly demanded de Gaulle's head, just as the F.L.N. today demands mine. From now on, I shall keep silent." This was the final phrase of his declaration to the court, and for an instant nobody else spoke at all. Later, the prosecuting attorney requested and received the judge's permission to question Salan, of whom he first asked, "Are not your heart and spirit shaken by the facts that have been recalled and of which you stand accused?" The prosecutor waited, and then said, "No answer. Do you consider your crimes legitimized by your intentions, and by what claim do you find absolution and excuse? . . . You do not answer. Do you consider that, lacking excuses, these crimes could be extenuated by your reasons for committing them? . . . Obstinate silence. I do not wish to comment on it. There are no responses for us to comment on, so even less may we comment on these silences." This monologue of deep, earnest curiosity addressed to Salan by the official who will surely demand his death was the most dramatic incident of the trial's opening two days.

May 31

The eight-day trial in the Cour d'Assises of the terrorist leader, ex-General Raoul Salan, was peculiarly monotonous, considering who he was and what was at stake. But its last half hour, before midnight Thursday, rang a sudden change. The finale was grotesque, bringing such an unexpected somersaulting of values that it became horribly ludicrous, as if justice itself had been posed upside down. In that tense, silent, animal moment when the courtroom waited for the verdict of life or death, the presiding judge declared "in the name of the French people" that the Military High Tribunal had answered yes to Questions One, Two, Three, Four, and Five, which all affirmed the prisoner's guilt, and yes to Question Six, the most important, which—had he been heard to finish his phrase—conceded extenuating circumstances. His words were lost but their meaning was comprehended in the violent partisan uproar of surprise or joy over the fact that Salan's head had been saved. Those who felt only a sickening shock sat silent, as if stunned by disbelief, or robbed. French pandemonium filled the courtroom. It was led, like a personal triumph, by Salan's stentor-voiced chief lawyer, who bellowed to the bench *"Merci, ah, merci!"* and embraced the prisoner, who leaned down from the box white-faced, astonished. His frenzied devotees, standing at the back of the courtroom, screamed *"Algérie Française!"* and were joined by many black-robed Palais lawyers, filling the aisles as if for the last act in a theatre. Some voices rose in "Le Chant Africain," the O.A.S. song; then came, inevitably, an ensemble shouting of "La Marseillaise." During all this hurly-burly, the nine members of the bench sat blank-faced, as if themselves amazed at what their pondered decision had wrought. Only after their unceremonious departure was it learned that in substitution for death Salan had been sentenced to life in prison. In court, as in Algeria, he had provided the most chaotic scene ever known there.

The majority of French opinion has been shocked and disgusted by the weak Salan verdict, like a last straw after the Gaullist Republic's cautious evasion of forceful tactics against the renegade officers when there was still time—when these dangerous Army generals and mad colonels had not yet created their uncontrollable private Algerian civil war. Ironically, de Gaulle was also greatly displeased with his Military High Tribunal for not giving Salan the death sentence, and over the weekend simply dissolved it (though it had taken his Presidential decree to set it up). Being nonexistent, it could not fulfill its special Monday task—and this is brutal irony indeed—of trying the youths who, one night last September, had attempted to assassinate de Gaulle on his road home to Colombey. This would seem an unconscionably long wait before getting around to trying would-be assassins of the head of state.

The public's great unsatisfied curiosity following Salan's trial lay in wondering what in the name of God—a name more than once mentioned during the last of the proceedings—could have served the Tribunal as extenuating circumstances. Legal circles say that it may well have been the brilliant bombshell that Maître Tixier-Vignancour launched on the eighth afternoon, when he theatrically declared, "The Parliamentary commission considering the amnesty bill met yesterday. Think, Messieurs, of your drama of conscience if, a week after General Salan has fallen before a French firing squad, an amnesty law is passed. For the rest of your lives you could not wipe away your remorse!" It is also conceivable that Salan's silence during the trial oddly served to save him. By his speechlessness, he became somehow partly absent; his muteness protected him, in a curious way, from intimate identification with what was said against him, as if, being dumb, he was also deaf and beyond reach. In contrast to the Jouhaud trial, no prosecution witnesses were presented who were still crippled from the O.A.S. sidewalk murderers' guns. Among the pro-Salan witnesses in plenitude there was signally the influential Mme. la Maréchale de Lattre de Tassigny, with her loyal, meandering reminiscences of her husband fifteen years ago in Indo-China, which she somehow draped around Salan, then at de Tassigny's side—the mandarinlike, already untalkative youngish French general of top standing. A pair of big-brass witnesses also spoke endearingly of Salan's patriotic past, bringing it nicely to life. It was as if Salan's career had been arbitrarily cut in two, and only Part I, which was estimable, was used in court, whereas Part II, bloodied over in this past year alone by more than two thousand murders in Algeria, mostly of Moslems, committed under his command, was never featured. Another omission was that in neither the Jouhaud nor the Salan trial was it ever mentioned what the Algerian natives were fighting for—their independence. Both were modern political trials that willfully omitted contemporary history.

The most important witness was de Gaulle's former Prime Minister, Michel Debré, who in defending himself and de Gaulle even defended Salan, in a way. To the outlandish suggestion (nor was it new) that Debré shared a moral responsibility for Salan and the other insurgent generals by having written, in 1957, as a rather obscure opposition senator in the Fourth Republic, that insurrection for *Algérie Française* was legitimate—as, indeed, it then was—Debré said, nostalgically, candidly, "*Algérie Française!* Who has not hoped for it over the last twenty years?" That it had profoundly "troubled consciences" and caused "bitter difficulties" was self-evident, he said, looking over the court. "*Hélas,* the world has evolved," he went on, decolonization having achieved such force and value that nations resisting it found themselves outside the pale. With Morocco and Tunisia having been given independent status before de Gaulle "was charged with the national destiny in 1958," Debré said, he decided in 1959 that Algeria's self-determination was legitimate and necessary, and in 1961 the French people's massive referendum made it law. This mention of the referendum aroused such a protesting mutter among the pro-Salan standees that the presiding judge threatened to clear the court. "If only one could do that in Parliament," Debré was heard to comment. After the cease-fire, he said, only two paths in Algeria were possible—"the road to reason and the road to folly." Looking coldly at the prisoner, he concluded, "The latter has led to blind criminal terrorism." But Debré's admission that the preservation of *Algérie Française* had been a deep emotional desire and a bitter problem had benefited Salan. He benefited, too, from the testimony of a certain debonair young captain, seven years on combat duty in Algeria, who said explosively of the F.L.N. *fellagha,* "He is a man who cuts everything—trees, roads, a nose, ears, hands, heads." Toward the end of that last tense day in court, Salan's secondary defense lawyer said, in his final, ultra-emotional speech, "It would be impious to take your life. If you must perish, you will mount the cross. But it is we who will be crucified. I salute you, *mon général,* and I cry to you *'Merci, mon général! Adieu, mon général!'* " The prosecuting attorney (who was suffering from a bad bout of sciatica and remained seated during most of his nearly two-hour final address, formally rising only at the end) with refined euphemism did not mention the word "death" in demanding it but referred to it as "the only irreversible punishment." His last words to the prisoner were regarded by many listeners as dubious theology and also as clearly ineffectual legal oratory, since they failed of their aim. Having pleaded with Salan to speak, if only to express his repentance, the prosecutor ended thus: "Do you not fear that, when your hour comes, God Himself, before your unremitting obstinacy—that God Himself will not deign to wipe the tears from your eyes?"

June 13

This is the final fortnight before France's seven-and-a-half-year war with Algeria comes to its factual end and solution. What was so long going on in blood will now finish up on paper, in ballots. On July 1st, the great day, the Algerian electorate—both the dominant Arabs and the minority whites (continually diminished by the daily thousands fleeing in panic to France)—will vote in the self-determination referendum on what form their future Algeria is going to take. That is to say, they will give their answer, which is sure to be affirmative, to the single question of vital interest to President de Gaulle, which constitutes his referendum's entire program: "Do you wish Algeria to become an independent state, coöperating with France under the conditions defined in the declarations of March 19, 1962?"—meaning the Evian accords. The Moslem masses being mostly illiterate, few of them can have read the Evian agreement, but they all know about the desirability of becoming an independent state, which is what their men in the F.L.N. were fighting and dying for all that time, and the answer will be yes, we wish it—even with France tied like a tricolor tail to their high, politically ambitious kite. The public announcement last Sunday of the single question marked the official opening of the referendum campaign. Special itinerant voting bureaus are to be set up in the oasis *départements* in the Sahara, where the population is small and scattered—the only *départements* where women vote. By Saharan tradition, the Tuareg women are the tribe's militants and go unveiled, and now that there is a ballot, they will use it. The ballots come in two colors—white for "Yes" and rose for "No," which seems an odd psychological *gaffe* on the part of de Gaulle's election agents. The white ballots have "*oui*" printed on them, and also transliterations of "*Kab-el*" and "*Kag-lah,*" which are Arabic and Kabyle, respectively, for "Yes."

Half a dozen political parties—most of them exclusively Moslem and all of them favoring a "Yes" vote—are active in the referendum campaign, the dominant one being the Front de la Libération Nationale, whose F.L.N. army made the war. Then there is the Parti du Peuple Algérien, led by the picturesque old Messali Hadj, formerly a Communist, now a Trotskyite, and the revolutionary founder, years ago, of the Mouvement National Algérien, the first ever organized to demand liberty, for which he was eventually arrested by the French and until recently kept in *résidence surveillée* on the lonely Breton Ile d'Aix. He was politically eliminated during the Algerian war by the new young bourgeois leaders, Mohammed Ben Bella and Benyoussef Ben Khedda, and is now rather a sad elderly nostalgic figure but still of some influence. The French

Communists and Socialists have each installed a campaign party, so as to have a finger in the Arab pie. There is no all-European party, which de Gaulle has stated he regrets, because it leaves incomplete the total representation he so clings to as a fetish of democracy.

June 26

Rosamond and Georges Bernier, editors of *L'Œil,* most intelligent and beautifully edited monthly art review here, have opened under its name an elegant new picture gallery in a huge, romantic Left Bank mansion at 3 Rue Séguier, around the corner from Picasso's atelier in the Rue des Grands Augustins. *L'Œil's* initial exhibition is a tribute to another art magazine, the famous, influential *Minotaure,* of the Paris nineteen-thirties. It was unique as the avant-garde progenitor and purveyor of the ideas of the Surrealists, then in their second and matured period—that militant group which, as you may recall, for the two decades between the wars was the strongest, most stimulating power combine in the French artistic field of new painting and writing. *Minotaure* served as a luxurious illustrated almanac—as often as there was enough cash to print it, which meant irregularly; in its nearly seven years of existence (from 1933 into 1939) there were only eleven issues, all bibliophile items today. Its founders were André Breton, Surrealism's Pope; the poet Paul Eluard; and two leading art-book publishers of Paris—the Swiss Albert Skira and the Greek Tériade. *L'Œil's* exhibition is made up of twenty-nine paintings and some sculpture, all of which were either reproduced in *Minotaure* or created during its regime by artists who were working Surrealistically at the time—by Picasso, Max Ernst, Tanguy, Arp, Miró, Chirico, Dali, Duchamp, Man Ray, Magritte, Brancusi, Giacometti, Masson, and Matta. Seen all together, they make a rare, historic show. Dali, of course, was expelled from the group as he approached the level of department-store decoration; Chirico left it for his private medieval limbo; Picasso moved on to his production of monsters; and Duchamp ceased painting. Aragon, the group's literary chief and the bitter rival of Breton, departed from its discipline to become a Communist Party *éminence grise;* one or two Surrealist writers were suicides and those of its major disciples still alive today are either venerated authorities or aging artists of international repute. In the gallery's foyer is displayed a row of medallion photographs of their faces when they and Surrealism were young, in the early twenties—faces of vivid, youthful heretics, with the heavy-jowled Breton already looking like a pope in exile, Dali like a picaresque Spanish male beauty, Ernst like an early Dürer drawing.

There are four Picasso contributions to the exhibition, all un-Surrealistic in style—nor is it sure that he was ever a Surrealist at all. Miss Gertrude Stein once authoritatively said, "Surrealism was no help to him." There is a grandiose, virile pastel, from his mythological period of 1933, of a minotaur crouching over a nude goddess, and also his "Portrait de Lee Miller" (now in the London collection of her husband, Roland Penrose), with her features randomly scattered about her face and yet with her identifiable 1937 likeness as the Left Bank American beauty invincibly held captive by his genius. A Brancusi bronze "Bird in Space" is on view, lent by Baron Philippe de Rothschild, who had never heard of the sculptor when he bought it but, as a racing-car enthusiast, responded to its suggestion of speed. The three Max Ernst canvases—the gigantic "Paradis," "La Joie de Vivre," and "Les Jardins des Hespérides"—demonstrate his superior aptitude for Surrealist art: his power of painting like a poet; his mastery of enigma, of the debris of dreams; his modest eroticism (libertinage being one of Surrealism's tenets); and his deep attachment to nature and its mimeticism.

Surrealism is démodé now, but it has left visible marks on French culture and American advertising. It was *au fond* a literary movement that aimed at paralleling itself in art—a doubled intimacy unknown here before, even in the time of the Symbolists. Created around 1924 by Breton, it has been his fame and will be his epitaph. Under his guidance, it seized on elements of the new century, which it aggressively popularized among the intelligentsia, rich or poor, if only by the scandals of its brawls in favor of all that it chose to be interested in. This included the interpretation of Freud, whom Breton personally knew; the sociology of Trotsky, another friend; automatic writing; oneirology; the irrational, the unconscious, and the psychosomatic; hypnotism, hallucinations, and free association; and ethnology as almost an esoteric study. Most of these are conversation pieces everywhere today, but they were not then. The dead writers the Surrealists were closest to were the Marquis de Sade, Rimbaud, Gérard de Nerval (who hanged himself in his top hat), and the so-called Comte de Lautréamont, who was their fetish for his malefic, imaginative creation "Les Chants de Maldoror." An acute international art critic has just commented, "The chief tenet of Surrealism was that contemporary art must be subversive, menacing, and a shock." Though the Surrealists were regarded by the French bourgeoisie as anarchists, they operated as prophets, for prophecy is what most modern art over the past forty years has turned out to be, even when not Surrealistic.

Brittany is now going through its second artichoke war, Artichoke War No. 1 having taken place about this time last year. The plethora of artichokes around St.-Pol-de-Léon, in Finistère, where they are tenderer and have bigger hearts and less spiky leaves than those raised elsewhere, has once more led to surplus, bitterness, a *jacquerie* among the farmers, and strikes among packers at the railroad stations, with some peasants dumping their artichokes by the thousand on the village streets rather than sell them below the established price. What is worse, the discarded chokes have been used as vegetable grenades in the resulting street fights, producing bloody rural faces, calls for the riot police, and official appeals for calm by the Ministry of Agriculture. In nearby towns like Brest and in the regional cigarette factories, tons of the surplus vegetables were offered last week to workers and townspeople for nothing, but they aroused hostility rather than gratitude. One smart worker said to a Paris reporter, "The St.-Polliens have the air of offering us charity. In reality, the chokes they hand out don't cost them anything, because they are subsidized by the government. So it's only a way to unload their stuff, and we taxpayers are the ones who pay." Last year, some members of the Primeuristes Indépendants, or the Independent Growers of Early Vegetables, trucked their artichokes down to Paris and sold them at neighborhood street markets with the old hawker cry of *"Voilà mes artichauts, tendres et beaux!"* Last Wednesday, their more powerful rival, the Société d'Intérêt Collectif Agricole, cut down the telephone poles outside St.-Pol, Plouescat, Plouvorn, and other villages, and laid them across the roads to prevent anybody from bringing his glut of artichokes to town to sell at less than the minimum price of forty centimes a kilo—about four American pennies a pound. The Primeuristes Indépendants, who had apparently planned to do just that, then telegraphed President de Gaulle, declaring that they "placed themselves under his high authority" to preserve their safety, "imperilled by the collectif's reign of terror."

Continued bloodshed, the blowing sky-high of buildings that represent and house civilization in cities, the miscarriages of justice, and the second thoughts about where justice is even to be found in a great country like France, which has seen the dignity of ending its own war removed from its control and given in part to men in prison—all these confusions, whether going on here or in Oran or Algiers, are like an amateur script for some kind of coarse, comic charade in which even anger, hatred, and desperation are the qualities of buffoons. The letter that ex-General Salan wrote a few days ago from his solitary-confinement cell in Fresnes Prison, advising the O.A.S. terrorist army he created and led to befriend the new Algerian independence, and begging the Algerian whites to stay by their cities and their land, largely ruined at his orders, was an ironic sequel to the earlier one from the same prison by ex-General Jouhaud, who offered the same solemn good advice—but to Salan himself, and at a moment when such good advice seemed a comrade's treachery. These elements have almost made judges out of the two criminals and traitors in finally settling France's war. Though it was France that the Arab rebels fought for seven years in pursuit of their independence, it is with the army of France's new enemy—the white French O.A.S., formerly high in France's own army—that the F.L.N. Arab forces have lately been agreeing to negotiate the peace. There is no common sense in the vital seriousness of all that has been going on. That the vote in the referendum to be held in a few days in Algeria will be overwhelmingly in the affirmative is right now the only seemingly certain consequence of this long colonial war, so often referred to by General de Gaulle as "*cette guerre absurde.*" The making of the peace has been permitted to be far more absurd.

July 10

The end at last of the Algerian war, on Sunday, July 1st, in the overwhelming self-determination vote in Algeria, and its transformation into official Algerian independence on Tuesday in Paris seemed almost a precipitation of history. It was the longest war that France has known in the West in modern times, and, during its last two or three years, the most generally despised by most of the French.

The only residents here who celebrated Tuesday with joy were the Algerian Arabs. After all, they had won—or, at any rate, had received what they had been fighting for. To honor their independence, they gave gigantic, orderly free feasts in their various Paris neighborhoods. In the poor Algerian quarter behind the Panthéon, in a humble Arab restaurant that normally serves a dozen *couscous* an evening, hundreds of plates were served free, like manna, from noon on, and any of us foreigners or French who had been the restaurant's clients were welcome. In the Rue de la Goutte-d'Or, the *médina* of Paris, just below Montmartre, the hospitality from noon on was more luxurious. This was the center for the gargantuan free victualling supplied by the Fédération de France du F.L.N. for its celebration—tons of semolina for the *couscous;* hundreds of sheep carcasses; hundreds of sacks of white and broad beans, of onions, tomatoes, and cucumbers for relishes, and of pastry flour for the *baklava* and fritters; and bales of mint for mint tea. Forty *médina* restaurants and cafés had been mobilized for the gratis hospitality (and the patriotic night of unpaid work in the kitchens and back courts that preceded the feasting), and for once the rather sorid Drop of Gold Street looked gay by night. There was no wildness of joy or shouting crowds on it—only the Arab men, dressed, if young, in their best French clothes, with all ages in a voluble state of disciplined excitement that was more impressive than noise. The tawdry Oriental arches over the café exteriors were hung with streamers and flags, and inside, behind the bars (serving free orangeade), among bouquets of greenery and more flags, patriotic mottoes were written on the fancy tiled walls in Arabic and sometimes in French, of which the most popular declared, "There is only one hero—the people," and a tragic one said, "Two million dead and an ocean of blood so Algeria might live free." Everywhere in the bars were amplifiers playing "La Marche des Moudjahidines" (the Arabic word for volunteers in the war), with its long, wavering musical phrases and flowery falsetto improvisations like garlands of Oriental melody. The most dramatic musical number—or so it was explained by a Kabyle acquaintance—was recorded in the Aurés Mountains when the men were about to go into battle. You hear them first called to prayer by the shrill voice of the mufti, because if they die as pious Moslems they will go straight to Mohammed's paradise; then comes a roll of machine-gun fire and, suddenly, the song itself—rhythmic, melancholy, and stirring, to a semi-French, Orientalized marching tune—which the Arabs in the bars each time joined in singing softly.

This was their new national anthem, composed in clandestinity.

The *pieds-noirs,* or white French Algerians, are still flowing in a panic tide across the Mediterranean to Marseille, in terror of possible retribution from the Algerian knives. It is this difference in the age of the two civilizations involved in the Algerian war that adds possible leftover horrors to the peace. When the war began, the majority of the impoverished, illiterate millions of Algerian natives were still at about the level of 1000 A.D., or the time of the Crusaders, with an inexpensive knife blade as their rational weapon, whereas the French Army's napalm fire bombs, bazookas, and high-powered guns achieved death at a distance, in the civilized twentieth-century manner. Actually, over the nearly eight years of fighting, the Arab peasants in the F.L.N. Army were understandably more modernized by their Czech guns and other Iron Curtain matériel than by the more than one hundred years of French occupation. More than a quarter million of the former million *colon* whites are now in France, and are being daily added to. They are a loss down there and not popular here.

September 12

The first major French film of the opening season stars Brigitte Bardot, directed, as in the beginning, by her inventor and first husband, Roger Vadim. It is called "Le Repos du Guerrier," or "Rest for the Warrior"—the title of a poignant, realistic first novel written a few years ago, apparently as a painful autobiography, by Christiane Rochefort. The book bore all the marks of tragic personal truth—the story of a young Parisian bourgeoise who goes to a provincial town to claim a modest inherited property; puts up at an inn, where by accident she opens the wrong bedroom door and discovers a young man dying as a suicide from an overdose of drugs; is responsible for saving his life, that of a sadist and drunkard; and is thereafter dragged down with him in the course of love. The arterial lifeblood of this novel Vadim has professionally set flowing on film in color. It displays Bardot as what she was born—a member of the well-off new Parisian bourgeoisie, though devoid of its official cliché of good manners—and no longer the exhibitionistic pretty rebel of her early film days. This new film does show her nude, though—three or four times—with beautiful, dignified photography.

In her new manner, Bardot is truculent, for in this film she is also very rich. There are brief moments in a series of bedroom scenes in which, in physical psychology and under excellent Vadim direction, she becomes the complete, competent little actress.

It was undeniably a shock last week for French radio listeners to hear the unmistakable Presidential baritone of General de Gaulle speaking to the massed citizens of Bonn during the first stage of his stately visit to Federal Germany, and saying, *"Wie sollte ich nicht bis in die tiefste Seele verspüren, wie bedeutungsvoll und ergreifend meine Anwesenheit . . ."* ("How should I not feel in the depths of my soul," *und so weiter*)—adding, for good measure, *"Es lebe die deutsch-französische Freundschaft!"* ("Long live German-French friendship!") Parisians seemed to have been unaware that he spoke German at all. His incredible mnemonic feat, at his age, of committing to memory his speeches *auf Deutsch* soon aroused a pride here in his excellently cultured and educated French brain that was as acute as the irritation aroused by much of what he said, when it was translated so his compatriots could understand it. His speech to the officers of the War College in Hamburg (and this one was in French) was considered outrageous here—that he should have said, with a high Army man's complacency (and, indeed, with a historian's accuracy), that the Germans "had never accomplished great things without something military having eminently participated in it." To the workmen at the vast Thyssen steelworks he said, again in German, "I wanted to come to you here where you work to give you friendly greetings from the French. For Charles de Gaulle to be here, and for you to give him so cordial and moving a welcome, is proof that confidence really exists between our two peoples." (The workers, it seems, unaccustomed to his protocolar language, were astonished to hear him refer to himself in the third person, as if he were absent, instead of standing there before them—so tall, one young workman said, that it was as if "the Eiffel Tower were visiting us.") The emotional, complimentary tone of all he said to and about the Germans, in his references to brotherhood and profound admiration, also offended many French. One Paris paper scathingly referred to his German trip as "Operation Seduction." An English correspondent in Paris said of his sentiments that "no other Frenchman would have dared employ such language." It must be added that at first the German population and press took him with a grain of salt;

the *Süddeutsche Zeitung* printed a delightful cartoon of old Adenauer in a top hat hanging an enormous welcoming wreath of laurel on the end of de Gaulle's enormous nose. At the last of his visit, however, the Germans were in a state of mob acclaim for him, reminding one cynical Bavarian of the response that Hitler formerly aroused.

What *Le Monde* had to say on de Gaulle's return to Paris was what French public opinion was waiting for, nor was it very generous. "Only the poor in spirit," it began, "could fail to rejoice in General de Gaulle's journey through Germany. That theatricality has played its part is not surprising. Weaned for years from spectacular political manifestations, the German crowds applauded a great and prodigious actor, and if they have to have a hero, even for a day, better a French general than a Bavarian corporal. In any case, through de Gaulle's person, it is the French people who, despite themselves, feel flattered and satisfied. All this is pleasant, sympathetic, and fragile," the paper went on with sharp disdain. The French people themselves, after three wars with the Germans in the past century, and two defeats, are immeasurably relieved that at last these two remarkable old men—or perhaps only their own astonishing septuagenarian—have tried to bury the hatchet for our time, which seems forever, the French people being convinced that though Europe cannot be constructed on two nations alone, it cannot be reconstructed without France and Germany in amity.

Owing to the second attempt on de Gaulle's life just before he went to Germany, and to the fact that nobody lifted a hand against him there (where he was discreetly accompanied, it was reported, by a German medical-corps unit carrying blood of his type, just in case), the French, on his return safe and sound, and after what he had accomplished, are saying, as they so repeatedly are driven to say, how difficult it would be to do without him—and, indeed, how hard it is to get along with him, too. This last new worry refers, naturally, to his announced project of assuring his successor, whoever he may be, to the governing of France by a vote of universal suffrage—preceded, of course, by the customary referendum to validate his project in the first place. Once again he plans to bypass Parliament, this time even to the point of reorganizing the government itself, and once again Parliament has declared that it will fight for its old rights, which none of its constituents fancy it will.

The immediate concentration of gratitude over the fact that de

Gaulle has not been assassinated as yet had a focus last week in the assizes court in the town of Troyes, where those would-be assassins who arranged to blow up his car at Pont-sur-Seine last September while he was passing on his way home to Colombey were being tried. There were many peculiarities about this trial, in which the leader of the murderous band—a certain Henri Manoury, former insurance salesman—had his head saved by Maître Jean-Louis Tixier-Vignancour, who also saved ex-General Salan's head, and who, being a rabid anti-Gaullist himself, makes a specialty of using his thespian gifts and Machiavellian legal talents as defense lawyer for such subversive criminals. He saved Manoury by insinuating that three members of de Gaulle's official Elysée family of functionaries had secretly been forewarned of the plot—if they were not a party to it. So susceptible are average French people now to suspicion of corruption and treachery in high places that the jury was influenced to the extent of leaving Manoury's head on his shoulders and giving him incarceration for only twenty years. And so great was the heat of the trial during its last days that when a witness happened to mention an outrage suffered at the same time by the French consul-general in Algiers, who, it seems, was undressed in public and beaten, Tixier-Vignancour insolently declared, "The posterior of a consul-general is never the symbol of France but is indeed the symbol of the present regime." He was not ordered from court or asked to apologize. During the Troyes trial, most of the would-be assassins who made the second attempt on the General's life, on August 22nd—this time by shooting machine guns at him (and Mme. de Gaulle) at Petit-Clamart while his car was once more en route to Colombey and home—were arrested, giving an extra, unneeded, shocking fillip to the melodramas of French justice and governing today.

September 26

Watching and listening to President de Gaulle on television last Thursday evening, as he instructed his nation on the special system by which he wishes his eventual successor to be elected and to function, one found it impossible, even three years in advance of that event, not to pity the incoming new President of the Republic in 1965, whoever he may be—floating out of his depth in the historic, turbulent wake that will be left behind by the unique, iconoclastic, enormous figure of *le grand Charles,* to whom the newcomer will be inevitably compared, if only as a form of intense relief to the anti-Gaullist minority and to practically all the politicians of France. For his TV speech, possibly his most important single selling talk since he took power, de Gaulle was in extra-good form, the mobile, elderly, unfatigued thespian face and the bold, inventive, ripe mind both seeming refreshed by his recent triumphal German outing. What he was pressing the French nation to accept was what he originally called (as you may remember) the American Presidential system of government, now become his own idea, to which he has just added our American system of electing our Presidents by universal suffrage instead of by an autonomous body—uninstructed by the voters—such as elected de Gaulle himself in 1958. These innovations, if made permanent, as he plans, would completely transform the entire political life of contemporary France. Both these American methods could be initiated, de Gaulle claims, by a referendum of the people, and without Parliament's assent, through the famous Article XI, on the organization of public powers, in de Gaulle's own made-to-measure, supposedly rigid and foolproof constitution, admittedly one of the best-drawn-up that modern France has had. In outrage, all the political parties (except, naturally, his own loyal, rather meek group, the U.N.R.) have unanimously declared that what he proposes would be a grave constitutional violation—one of the few acts that French politicians regard as heinous. Furthermore, de Gaulle's plan for an Americanized President would make the President so far superior in power to the Parliament that Parliament would perhaps be rated even lower in value in the future than it is now, under de Gaulle himself, whose high notions of the American President's supremacy over Congress seem in many ways to have little to do with the realities in our White House.

De Gaulle's harsh comments on Thursday against French politics and their politicians did nothing to soothe the latter. His optimistic determination to further rescue his beloved France by leaving it with something more solid in the way of a future government than the short-lived, dying-duck, Parliament-run governments of the past logically forced him (and how he relished it!) to enumerate the appalling situations they used to lead to—"the chronic confusion and perpetual crises" that periodically paralyzed the country, and "the abyss awaiting France if, unhappily, it were to fall anew into the sterile, ludicrous political antics of yesteryear." He also deplored France's curse of political divisionism, meaning the half-dozen or more political parties demanded by Frenchmen's fundamental inability to agree with each other on much of anything, to which de Gaulle himself has added three more divisions—those who revere him as a savior, those who do not trust him because of his autocratic Caesarism, and those, alas, who so hate him that they try to assassinate him.

The blast that his speech received from the major political parties, the political leaders, and the newspapers seemed the most united and hostile ever directed at him since he became President. Parliament's ancient, most authoritative deputy, the tiny Paul Reynaud, of the Indépendants, unexpectedly said of de Gaulle's projects, "Government by President works badly in the United States and would work much worse in France." Maurice Faure, leader of the Radicals, said, "No jurist in the world would swallow de Gaulle's interpretation of his own constitution." The Communist *Humanité* imaginatively declared that his scheme for a popularly elected Presidential government amounted to "a revival of the monarchic principle of divine right." The Socialist journal, *Le Populaire,* ironically jeered, "Long live personal power!" *Combat,* the morning paper of the intelligentsia, said that his election project laid the basis for a *"monocratique"* regime, and the ever-influential, stately *Le Monde* feared an excess of *"monocratie,"* neither of these words being listed in the new and remarkably up-to-date 1962 Petit Larousse dictionary, though at this point in de Gaulle's career, apparently, they should be.

October 4

Now that Parisians are home from their vacations, they can enjoy fine sights they never saw before in the middle of their own city. The *blanchissage,* or façade-washing, of the major elderly historic buildings radiating from the Louvre, which was begun in June, 1961, as a five-year program of aesthetics, by M. André Malraux, State Minister of Cultural Affairs, is now well along its course. The result is superb, at last popular even with the ordinary public, which feared that the sense of French history would be washed away with the dirt. Cleanliness has restored the architectural youth of these majestic piles, and one sees them in their original fresh, pale sixteenth- to eighteenth-century grandeur, as if one had the ocular privilege of being several hundred years old. This summer's really glorious revelation via soap and water was the intricate beauties, previously black and unintelligible through time, of the Cour Carrée, that huge square inner court of the Louvre, which few tourists—footsore from the picture galleries—ever have the strength to walk back over the cobblestones to view. It is the greatest art exhibition in Paris right now, and should be so advertised this winter and the next few springs and summers. (It will stay clean a longish time, being sheltered from the motor fumes of the streets outside and being prohibited to cars.) The chef-d'œuvre of the French Renaissance, it was in part carved by the great sculptor Jean Goujon himself, and its pristine complex incised beauties are now clearly legible on the three-story, almost blanched stone walls—a carved mixture of pagan gods, statuesque goddesses as caryatids, bearded Greek philosophers, fat cupids, and flower garlands, with the ciphers and initials of all the kings for whom the Louvre was built (from François I and Henri II through Louis XIV) visible amid rows of pilastered Corinthian columns. The famous, elegant Perrault outside portico, across from the church of St.-Germain-l'Auxerrois, has been cleaned, and so has the Madeleine. Its dirty dignity gone, it now looks attractive, even though cleaners in green oilskins are still scrubbing, with hoses and little rags, the last traces of soot embedded in the runnelled columns, and its distant companion piece, the Palais-Bourbon, or Parliament, is as spotless as if it had never known a political smudge. Other masterpiece buildings cleaned are Mazarin's curvaceous Institut and Louis XIV's Invalides, with its handsome carved ornaments of war and armory such as had crippled the invalids living inside it, and now Richelieu's Palais-Royal garden walls are undergoing the treatment, driving the ordinarily lucky inmates of the coveted apartments mad with flapping rubber curtains, scaffoldings, dripping water, and the omnipresent oilskinned men. Most of these great buildings belong to the state or to the Académie des Beaux-Arts, though there is some private ownership on the Place Vendôme (also being cleaned), where the Morgan Guaranty Trust Company has led the way. What the state, the Beaux-Arts, and the private owners have in common is the high price of the cleaning—nine New Francs, or a dollar-eighty, a square metre for plain soap, water, and scrubbing-brush treatment, and at least thirteen New Francs for cleaning by detergents, which kill the stone disease. Whether Notre-Dame can and will be cleaned—a rumor that has caused considerable excitement in Paris—remains for the ecclesiastics to decide, it being Church property. Laymen seem to think that the job needed for Gothic and gargoyle would take an infinity of time and would cost more than the Church would think it prudent to pay. After all, Notre-Dame has been famous, dirty, and inspiring, for nearly a thousand years.

October 10

It seemed characteristic—and certainly it was an old familiar act—that the first use the French Parliament made last week of its suddenly unloosed political energies after these four inert years under General de Gaulle's autocratic Fifth Republic was to overthrow a government—his government, headed by his Premier, Georges Pompidou. This opening and successful defiance of the all-powerful President de Gaulle took place in that dramatic, historic all-night Parliamentary session which ended shortly before dawn last Friday. Its damaging and, at the same time, liberating results will be spread over the next month and a half, until a French government is put together again, with a newly elected Parliament. But whatever comes now will be only like a postscript to the heroic Fifth Republic.

As harsh, realistic proof that de Gaulle was indeed the savior of France, his sacrosanct salvationist importance has ebbed in the few months of peace since he ended the Algerian war. His solo pattern of governing today's prosperous, revitalized France has seemed increasingly démodé to many of the French. This, combined with his age and the recent appalling and so nearly successful attempts to assassinate him, made imminent the French Parliament's return to the surface of national life, and, unfortunately, the inevitable restoration of some of the bellicose republican practices and politics normal to France—provided that Parliament could find an opening, some chink or hole, in de Gaulle's impervious personal armor for its reëntry. It was this that he supplied by a so-called violation of his own constitution—contemptuously bypassing Parliament and relying only on a coming popular referendum to validate his project for the election of his dynastic Presidential successors. These he proposed to have elected by popular suffrage, in order that they might not be selected by Parliament's politicians. De Gaulle's war against the French politicians has been like a religious war, in which the politicians a fortnight ago suddenly found him guilty of a kind of heresy—a violation of the Sacred Writ of Constitution. Here was their chance to attack him.

That de Gaulle had in truth violated his constitution was also the opinion of France's Council of State (which corresponds to our Supreme Court), of the majority of the Sorbonne law faculty, of leading jurists all over the country, of the powerful French trade unions, and of Paris and provincial newspaper editorialists. It was also the opinion of his two immediate predecessors, ex-Presidents Auriol and Coty, and of that high official who would follow him as a stopgap President if he should be murdered or should die in office—the notable French Guiana Negro M. Gaston Monnerville, long-time president of the French Senate.

All this disapprobation reverberated as cumulative news through the press from one end of France to the other. The only sign that de Gaulle had overheard it at his height was his astute decision to give his nationwide TV speech twice last Thursday—first at 1 P.M., before Parliament convened and started talking against him, and then at eight that night, his usual time. Relayed over the state-owned Radio-Télévision Française, it was one more demonstration of his unique privilege of communication, of personal propaganda, and of influence with his special little nation of eleven million radio owners and three million TV owners, to whom his gifted microphone voice and his dramatic, elderly face, now the visage of France herself, have become exclusive symbols of state leadership. One must understand that the power of his privilege is enhanced at all times by the R.-T.F.'s being a state monopoly and strictly censored, so that, in the ordinary way, no one and nothing subversive or antagonistic to or critical of the state—which means de Gaulle's government and himself—ever gets on the national French air. (However, this monopoly will be temporarily lifted on October 15th, the official opening of the referendum campaign, when his opposition can have its belated say.) De Gaulle's most important announcement in the one-o'clock broadcast, which certainly had some influence on the afternoon Chamber speeches, was his closing solemn threat to the French people that he would retire from their midst if they failed to support him adequately in his now truly vital

October 28th referendum on that Presidential-election project, as much a necessity for his plans for the future of France as it is a measure of his prestige.

All this was the agitating background of the extraordinary Thursday-into-Friday Parliamentary session that brought to the foreground this first organized political defiance of the monumental, solitary *chef d'état*—the first pandemonious, garrulous attack in what has openly become an intense power struggle between him and them, they being the long humiliated and now vengeful "men of politics," as he disdainfully calls them, as if somehow rather illegitimatizing them. The aim of the Thursday session was to pass a motion of censure against de Gaulle's violation of the constitution to carry out his Presidential project as an "opening breach through which, someday, an adventurer might pass"—a remarkably pretty and sinister phrase. He himself was, naturally, absent. At the session's beginning, the deputy-filled hemicycle hummed with revived animation. Once more, Parliament was the disputatious center of Paris, of France. A choice audience was also there assembled, almost as many women as men, drawn by a special appreciation of the unrehearsed, ad-lib drama, in which they could watch and listen to a government that may end in mortal agony among its final and futile political forms of speech—an audience that continuously packed the luxurious visitors' loges and crammed the public gallery's wooden benches beneath the roof during the twelve hours (from four o'clock Thursday afternoon until four-thirty the next morning) of intermittent speeches, debates, shouts, cheers, insults, interruptions, bangings of desk tops, and occasional long suspensions. Then the figures on the overwhelming vote of censure—more than three-quarters of the entire house being in favor—were read aloud, and de Gaulle's government was dead.

Another exceptional thing about this session was that the five traditional old big parties, all quarrelsomely opposed to each other in Parliament in the past, had this time united as an opposition. As their opposition leader they had chosen the Chamber's most brilliant, vivid, quick-tongued orator—the diminutive, sturdy octogenarian Paul Reynaud. Chief of the attack against de Gaulle on Thursday, thirty years ago he had been the first in political circles to discover and admire the then unknown Colonel de Gaulle, with his vain dream of a modern, mechanized French Army. As Premier of France in the early, desperate days of the last war, Reynaud had made de Gaulle Deputy Minister of War; in 1958 he had helped bring him back to power; and recently he had broken with him over republican principles when the General chose as his Premier the estimable Rothschild banker Pompidou, who had never even been in politics, let alone been elected a deputy. Reynaud's Thursday speech glittered with emotion and intelligent substance. Standing at his full brief height on the speaker's rostrum, and as if musing in disillusion over his illustrious former friend who had gradually seized all the functions of government in his own two large hands, he said, "How could we have fallen into such intellectual disorder? It is because General de Gaulle wanted to combine the honors of the Chief of State with the powers of a Premier—to be both Winston Churchill and George VI. . . . To those who say with fright, 'But what if he should leave us?' "—the threat contained in that noon's TV speech—"I say that this fright is not justified unless you doubt France. It is not very deeply patriotic to lose faith in all the French except one. To use this argument does not make one man bigger, it makes one's own country smaller. . . . In all civilized lands, the Parliament represents the nation, with its qualities and defects, its diversities, and even its contradictions. If the Assembly represents the nation, then France is here"—in the hemicycle itself—"and not elsewhere." His final, rather gallant sally was addressed to de Gaulle's Premier Pompidou, listening as if mesmerized on the government's front bench. To him little Reynaud ordered loudly, "*Monsieur le Premier Ministre,* go tell the Elysée that our admiration for the past is intact"—a noble compliment to de Gaulle's earlier days—"but that this Assembly is not degenerate enough to renounce the Republic!" There was then a burst of what the French papers called *"vifs applaudissements prolongés sur les nombreux bancs."*

A very noticeably high proportion of the thirty-eight speakers listed were from de Gaulle's own party, the U.N.R. These were mostly self-important, inexperienced young men with nothing consequential to say. However, the charge of favoritism was not made until fairly late in the night, when an angry deputy suddenly announced from the floor that the regular nine-thirty TV news—broadcast, of course, by the government and shown all over Paris—had given exactly one and a half minutes of its afternoon Chamber news to Reynaud and his dominant speech; the same to Socialist leader Guy Mollet, his colleague in opposition, who followed him; and a full thirty minutes to Premier Pompidou's exegesis of de Gaulle policies. At this, the Chamber broke into an uproar of boos, shouts, desk thumpings, and angry voices crying *"Voilà le fair play français!"* Pompidou himself seemed both surprised and embarrassed by the discrimination. The Speaker of the Chamber ordered a long suspension to consider what to do, then reported later, to repeated booing, that no TV news at all would be recorded of the evening session's proceedings, which increasingly looked like a defeat for the de Gaulle forces. The session being resumed at midnight, the passage of time was carefully indicated by a change in the attire of the Speaker, young Chaban-Delmas, of the U.N.R., who, sitting aloft in his great chair on the tribune, with his gavel ready to rap to restore order and a bell to clang if the noisy deputies started to get really out of hand, looked as he always does—rather like an expressionless, handsome tailor's dummy. For the afternoon opening, he had worn a very smart black cutaway coat and striped trousers. For the night session, he reappeared, as he invariably does, in full evening rig of tailcoat and a somewhat flamboyant big white butterfly tie. Just before dawn, when he announced the final news, he was probably the only man in the Chamber still immaculate and unwrinkled.

That portion of Pompidou's speech which was devoted to de Gaulle as a man, and was not a defense of his fashion of governing, was bold and touching. In part, he said, "This President of the Republic, General de Gaulle, is not a general in the popular sense that General Bonaparte or General Boulanger must have been. But it is he to whom you all, *Messieurs,* owe not only the restoration of the Republic in 1944 but the saving of the Republic in 1958, and again at the Algiers barricades in 1960, and again in 1961, during the *Putsch* there, and it is not six months since he surmounted the offensive of the O.A.S. Even on the evening of the recent assassination attempt, there was no one who did not feel that with him liberty nearly died. I beg of you at least to temper your words against him with gratitude."

When it was all over, the winning deputies left the Chamber shouting the "Marseillaise," with no unity of rhythm or pitch. In the courtyard outside on the *quai,* across from the illuminated spectacle of the Concorde's pale, cleansed, porticoed buildings, moon-colored just before dawn, a soldier stood in the shadows, with his cape on, his feet spraddled far apart, and his machine gun at the ready in his hands in case there might be the beginning of trouble because of this first decisive political crisis in de Gaulle's Fifth Republic.

The painful news, just announced here, of Sylvia Beach's death sometime last week, alone and for days undiscovered in her small flat above what had been the Shakespeare bookshop premises, is another American epitaph to engrave on the historic Left Bank Epoch of the Twenties, to which she was the hard-pressed presiding hostess, book publisher, book lender, bookseller, and friend, who never failed in any line. Without her, some of those who were becoming great writers could not have written so well, and certainly without her James Joyce's "Ulysses" would not have been published as early as it was. Her eyesight had never been good, and deciphering Joyce's curlicues of crabbed handwriting and interlinear scribblings on his manuscripts and proofs was an exercise of devotion and loyalty that her eyes never recovered from. She was a friend to all of us in Paris who used the pen, no matter how modestly. She was a friend to writing.

October 24

France's present crisis in government seems malapropos. It is a moment when no trouble at all in France would be welcome news. Actually, France's brief and bloodless current difficulties—or at least the first part of them—should be settled over the weekend. On Sunday, the French are to vote yes or no in President de Gaulle's referendum (which proposes the direct election of future Presidents). But this referendum has in the last ten days unfortunately turned into something quite different. It has been transformed into a bitter, explosive plebiscite on the popularity and the merits of de Gaulle himself, and into a violent fight against him personally, which has been nicknamed *La Bataille du Non*. It was launched and is being carried on by the five traditional main political parties of France, temporarily united in this effort of destruction, an all-out national attack of criticism, derision, and even insult against him—by word of mouth, by print and press, and by public meetings—with the politicians battling against him both as a legend and as a leader, and showing a frustrated fury and pent-up hatred that they have never dared to show until now, and whose revealing violence has been a startling surprise. What the politicians actually aim at is to denude him of his glory, to push de Gaulle askew on his pedestal, to drape him in ridicule and precisely laid-on criticism so voluminous that his image can never again seem the same to the voters of France. This iconoclasm is certainly a not unintelligent part of their power struggle going on here, and is due to come to its real head in the November elections of deputies to the new Parliament. This new Parliament cannot be dissolved by de Gaulle for one year, so it is of acute importance politically. What the politicians quite naturally want is to fill this Parliament—to cram it, if they can—with anti-de Gaulle deputies, who will continue in the hemicycle the power struggle over who will govern France: he or they.

To the voters, the de Gaulle referendums are like personal dialogues between him and them, in which he asks them to give him something he especially wants, and over four years they have developed a majority habit of giving it to him, loyally. The only thing that could make the Sunday referendum very serious indeed would be a failure of the affirmative majority to be massive enough to satisfy de Gaulle's present need for increased prestige. In this case, as he candidly threatened in his national broadcast last week, he would at once quit the public scene, never to return from his retirement at, supposedly, Colombey-les-Deux-Eglises.

Either the Gaullist regime or France's outworn but still ambitious political-party structure, now nearly a hundred years old, is bound to be seriously weakened in the next four weeks. What may happen is that the people, out of their proved personal devotion to de Gaulle, will vote this Sunday to keep him in power, and then next month, out of their inherited political affiliations, will elect a strong opposition majority against him in Parliament. Once again, de Gaulle might be able, as he has been over the past four years, to keep them under wraps and quiescent—to dominate them at his distance—but with greater human difficulties this time. He may remain in the saddle, but they will be in rebellion.

In the reports on the Battle of No in the daily Paris press, de Gaulle's detractors have angrily defined him as, variously, a megalomaniac, an egomaniac, a tyrant, a suborner of justice (since he has not obeyed his own decrees), a violator of his own constitution and therefore an illegitimatist, a leader who stuffs the public's head with nonsense, a demagogic flatterer, a strangler of the Republic, an adventurer, an absolutist, an autocrat, a misanthrope, a dictator, and a camouflaged monarch. M. Daniel Mayer, a leading Socialist, has just stated, "De Gaulle risks becoming the von Hindenburg of France." *L'Humanité,* citing his last TV speech, in which he threatened to leave if not supported, riposted with "The sooner the better," as the title of its answering editorial. The morning *Figaro,* though it prints certain polite anti-Gaullisms as part of today's normal news, has tried to remain editorially quasi-loyal to him, but wishes that he had not made so much trouble for himself and everybody else by setting the yes and no French at each other's throat. *Le Monde* also treats him rather strictly. In its Saturday editorial, it frankly said that the yes and no were like Scylla and Charybdis, and that everything depended on whether or not the new Parliament would set up a modern Republic "after the passage of de Gaulle's bulldozer," meaning his referendum. It then added, in insistence, "The referendum's true problem is not to save de Gaulle or to throw him out but to determine the future of the Republic. The worst consequence of a no vote would not be the departure of the General but the inevitable return to the traditional Parliamentary regime—to the certain restoration of the Fourth Republic." The only purely pro-de Gaulle paper in all Paris is the limp, one-page daily *Nation,* operated by the General's rather browbeaten political party, the Union of the New Republic. All the close to a dozen other Paris papers are, if not openly against him, at least not devotedly or continuously for him these days—not by a long shot.

November 7

It is not too late to mention how Paris reacted to the alarming world crisis between President Kennedy and Chairman Khrushchev of two weeks ago, because the Parisians themselves have been so late in putting their minds on it that they have expressed their full opinions only over the last few days. There were, it is true, some earlier snap judgments by the conventionally anti-American intellectual left-wing voices and periodicals here, shouting to the White House "Hands off Cuba!" and jeering at what they called Kennedy's obvious buildup of the whole affair as an electioneering move to help his party in this week's elections. Actually, the only President who used the Cuban tension in this fashion was President de Gaulle, who, in a final TV speech to his own electorate in his recent referendum campaign, alertly warned them that they were living in a dangerous world and had better vote for him massively, so he could take care of them. Indeed, the initial Paris political reaction to the Cuban crisis was a fear not that the whole world might blow up on Monday or Tuesday but that on Sunday Cuba would be a gift of the gods to de Gaulle's referendum, bringing him millions of otherwise wavering votes, which it probably did.

November 21

The single dominant figure in last Sunday's preliminary parliamentary elections, who was not even a candidate and yet won with a landslide of ballots all over France, was President Charles de Gaulle. Actually, it was, of course, his party, the Union pour la Nouvelle République, which, as the political go-between, received the millions of votes, but since they were meant strictly for him, they made him the utterly unexpected transcendent and spectacular winner. The U.N.R. received five and three-quarters million votes, or almost thirty-two per cent of all those cast—a startlingly high figure for France at any time, and especially now, considering the complexities of the competition. Of all the traditional old-line parties, only the Communists gained, and even the Communists—for a decade called *"le premier parti de France,"* because incomparably the most numerous—lost their title to the Union for the New Republic, now become France's first party.

According to the dazed Paris interpretation of all this, de Gaulle's victory can be attributed in part to his intelligent sense of provocation, which made of this election a national crisis, and to his brilliant, classic capacity for aloof, superior planning. Already it seems clear that he did not win the election on his legendary personal popularity alone, today worn thin by the pressures of ingratitude, legitimate criticism, and time. He won, in great part, on the unpopularity of Parliament. There is here a long, unabating antiparliamentarianism, a form of political non-belief held by millions of French citizens of all classes, lodged in their minds by memories of the shambles and national humiliations in governing that were perpetrated by parliamentary politicians during the inefficient, confused Fourth Republic—and the scandalous, wicked Third Republic, if their recollections go back that far. This week's shattering blow against Parliament—at least in its previous, too familiar form —was the result of a manifest lack of faith or interest or hope in it as the quasi-sacred political machinery of France. As soon as the election results were published on Monday morning, the Paris Bourse rose four per cent, and Switzerland cancelled its weekend selling orders.

December 6

As if in sudden recognition of the fact that there are more women in France than men (they outvoted the men seven to six in the recent parliamentary elections), for the first time in French publishing history three of the main year-end literary prizes, including the Prix Goncourt, have been given to women writers. The most stimulating literary criticism annually connected with the Goncourt Prize rarely concerns the novel that has just won it. The criticism is almost invariably directed against the Prix Goncourt itself as an institution—a vestigial, erratic, and powerful publicity enterprise that, merely by tradition, can once a year turn a book into the national best-seller. As preparation for the recent Goncourt Prize day, a round robin of critics publicly declared that no intellectuals ever read the Goncourt selection anyhow. In the literary, and even in the political, weeklies, critics vented their customary vexation at the Goncourt's basic weakness—the mysterious lack of critical acumen displayed by the Goncourt jurymen, themselves respected writers, playwrights, and academicians. In *Les Nouvelles Littéraires,* France's most notable literary critic, Pierre de Boisdeffre, sarcastically inquired, "When a writer of talent is given the Prix Goncourt, isn't it because of some misunderstanding? Have Gide, Mauriac, Giono, Montherlant, Saint-Exupéry, Sartc, and Camus been the losers in any way in never having been distinguished by this honor? The novels that in September start piling up"—three hundred is the average number submitted by their publishers for the various November prizes—"are not books but lottery tickets," in which luck, not literature, will win.

This new Prix Goncourt novel bears the bitter title "Les Bagages de Sable," or "Luggage Filled with Sand," and was written by a Polish émigré, Mme. Anna Langfus. Under the heading "A Charity Goncourt," one critic exceptionally and cruelly wrote of it and of her, "The story is poor, the writing is poor, and the author is also poor, no doubt." The Goncourt Prize automatically sells around a hundred and fifty thousand copies, bringing in royalties equivalent to about forty thousand dollars—perhaps the only cheerful item that can be associated with this painful and obviously truthful book. It concerns an impoverished Polish refugee in Paris, so scarred and sickened in her memory and body from the brutalities suffered during the war in Poland that even love, when offered to her—only by an elderly lover, it is true—fails to heal her.

With the opening today of the Fifth Republic's recently elected new Parliament, which contains, for practically the first time in the history of all of France's Republics, a majority party—de Gaulle's, of course—it is conceded here that France's republican system, as it has interruptedly been known for the past hundred and seventy years, has now come to a full stop. Something fundamental is changed in France. The three elements of French history that the French today still seem proudest of are Louis XIV, the French Revolution, and the fact that France is a republic. Yet the French have always experienced a great deal of difficulty living in and with their Republics, which so far have always turned into something else. The First Republic, of 1792, in twelve years turned into Napoleon's Empire. In 1848, the Second Republic, of almost four years, turned into another Bonaparte empire, enthusiastically voted for by a landslide of Frenchmen. In 1870, France began "going into the Third Republic backward," as the phrase then was—meaning reluctantly, since another monarchy had been hoped for by most leaders except Gambetta. The corrupt, tough Third Republic (nicknamed "La Gueuse," or "The Slut," and the longest regime that France has ever known since that of that great Louis XIV) ended in 1940 in the Vichy state of Maréchal Pétain, the only Frenchman who ever survived a hundred and seven French governments and founded one of his own, the hundred and eighth. In 1944 came de Gaulle's Provisional Government, and in 1946 came the Fourth Republic, which lasted twelve weak and addled years before it fell into de Gaulle's Fifth Republic, now four and a half years old. The general tenor of the current political commentators has been that the weakness of this country's Republics and Parliaments lies in the fatal gift of the French for individualism and for never agreeing with one another. As de Gaulle himself dryly said, "How can you govern a country that has two hundred and forty-six varieties of cheese?"

1963

January 2

The worst of the exceptional cold snap here, and all over Europe, exactly filled the holiday week from Christmas Eve through New Year's Eve, delaying letters, greeting cards, telegrams, and telephone calls, stalling buses and trains on frozen roads and rails, closing down airports crippled by icy runways, and somewhat freezing the travellers' seasonal spirit of joy and good will. Snow avalanches impeded the journeys of impatient skiers; snowslides closed the St.-Gotthard Tunnel, stranding thousands of home-going Italians. Italy had its worst cold wave in a hundred years. France had its own South and North Poles of cold: the southwestern part of the Côte d'Azur, centering on Marseille, which was snow-bound, and the northwestern slice of Brittany, where, as the French phrase it, *il a gelé partout à pierre fendre*—it froze hard enough all over the place to crack the very rocks. The canal that connects the Rhone and the Rhine was frozen between Strasbourg and Neuf-Brisach, and the waterway that runs from Belgium down into France was clogged with hundreds of barges of coal, which Paris needed, sitting paralyzed. The city had the coldest Christmas Day that it has ever known. To Paris motorists, little used to struggles with ice on forest highways or to skirmishes with snow, such weather seemed dramatic. Guests from the city arrived belated and extra-hungry for the midday festal turkey dinners in suburban country houses, where the bathroom pipes were mostly frozen. Because of thawing and refreezing, provincial roads became so dangerous for motoring that President de Gaulle, after his holiday at his house in Colombey-les-Deux-Eglises, which has no railway connection, returned to Paris in a special train on the nearby Troyes-Chaumont line. The ice on the ponds was thick enough to give young people happy days of skating, and the deer in the various forests had a quiet holiday, since hounds could not follow their scent where it had snowed, and the frozen ground was too dangerous for horses to jump or gallop on.

June 19

President de Gaulle is periodically still the most interesting Frenchman of all France to the French, and doubtless to himself as well, almost constantly. For nothing interests him like French history, and he is the only one with the power to continue making it. If one returns to France after a six-month absence, one finds that nothing consequential has changed in the interim except under his impulsion. He has all the causal elements between his large hands and within his broad imagination, and he represents the successful results. This last week, he made a vital segment of new history, which is not only French history but also European. After Parliament spent all last Thursday in often brilliant scathing debate, led by the minority old-guard Opposition—an opportunity for recrimination against de Gaulle, his deputies, and his policies such as the minority rarely has a chance at—the Chamber, balloting at midnight, gave a strong three-to-one vote (with the minority's unexpected assistance) for the ratification of de Gaulle's extraordinary Bonn treaty of Franco-German collaboration, which Parliament had been called together to consider. This treaty not only buries the hatchet on paper between France and Germany but invokes the establishment of Franco-German friendship like a new bridge across the Rhine. Many deputies and also many bourgeois sections of French society felt cold, suspicious, and grudging about offering even to the Western remnant of the former Nazi nation this Treaty of Coöperation, its official title. The former French Premier, old Guy Mollet, the Socialist Party Chief, opened the attack by caustically telling the Gaullist government bench, "Your only interest in Europe is in a balance of power—an English-style Europe without the English." Then de Gaulle's Prime Minister, Georges Pompidou, the former Rothschild banker who in a short time has become an alert, able Chamber debater, invoked the United States' Cuban crisis as proof that the free world's situation in the nineteen-sixties has changed. Pompidou explained that in the United States' recently revised defense program for herself and for her allies, the proportion seemed like that in the recipe for the legendary old-world lark pâté—one lark to one horse—and added, "I must say that a lark for our defense is not a sufficient guarantee." To American and English listeners present, Premier Pompidou's most dazing contribution to this public session, which aired the Opposition's many criticisms of the Franco-German coöperation treaty (although the deputies had no power either to change or to amend it), was his patient, courteous reminder to one captious anti-Gaullist deputy that, after all, "the government need not have submitted this treaty to Parliament. But it has done so." It is with such a sovereign free hand that President de Gaulle can create his continuing French history.

This year, the government's annual estival art attraction for Parisians and visitors is a centenary exhibition of the works of the prodigious Romantic painter Eugène Delacroix, who died here in the summer one hundred years ago. It is an extremely thorough collection. Five hundred and twenty-nine items, featuring his major grandiose pictures surrounded by their conceptual sketches, are on view in the Louvre's Salon Carré and Grande Gallerie (both badly lighted); his drawings are displayed in the Cabinet des Dessins; across town, his black-and-whites are on show in the Bibliothèque Nationale; the Parliament and Senate libraries are open to show his frescoes; his studio in the Place Furstenberg has been unlocked for sightseers; and a provincial exhibition of his paintings is in the Beaux-Arts at Bordeaux, where his father was a prefect under Napoleon, when the painter was a little boy. Because none but normal attention has been paid Delacroix for years, Paris cultural weeklies have gone into a Delacroix delirium. *Les Nouvelles Littéraires* featured him on thirteen of its vast pages, offering articles on Delacroix the Romantic painter, the writer, the dandy, the political revolutionary, and the colonial traveller; on his love affairs (which were few); on his paternity (apparently he was not the bastard of the famed Talleyrand, as was rumored by Mme. de Staël, but the son of his own dull, officious father); on his portraits and his battle scenes; and on his love for the music of his young friend Chopin, whose piano he had brought over to his own studio, so that the hypersensitive Pole could compose there in peace, away from the embraces of George Sand.

"My pictures achieve tension," Delacroix wrote. According to French analyses, he melted neoclassic formalism into light, subjected form to color, started the modern evolution in French painting that culminated in Monet, and affected Impressionism, Fauvism, van Gogh, and Cézanne. His compositions were too strong but glorious, whether depicting man, beast, or history—ancient or of his own time, for he was sensitive to both. It is these familiar history scenes that are the most popular and worthy in the Louvre show. He used the 1830 Paris revolution, which he saw but did not participate in, in his canvas called "July 28, Liberty Leading the People," with a top-hatted citizen on the barricades bearing a musket, and bare-breasted Liberty triumphantly holding the French flag. His sumptuous "Death of Sardanapalus," most grandiose of these anecdotal masterpieces (some of which need cleaning), with its amplitudes of scarlet, female flesh, and drama, was the scandal of the 1827 Paris Salon. When His Majesty Louis-Philippe, who bought Delacroix pictures for his royal gallery, sent the Comte de Morny to Morocco after the French conquest of Algiers, the artist accompanied him, and was thus the first French colonial painter—of caïds, of warriors in burnouses and sashes, of Arab horses. The odalisques that Matisse later imagined and painted in Paris, Delacroix painted in their corporeal reality in North Africa.

August 6

Since the first of this week, Paris has been what is annually called empty, about two million of its three million inhabitants being someplace else. August is like a form of restful paralysis for Paris, left with almost no motion: with nearly no traffic (including nearly no taxis) and nearly no modest restaurants open; with most grocers, butchers, bakers, and small shops of all sorts padlocked tight; with those minority Parisians who are still in residence seeming to remain indoors, as if under a permanent curfew; with parks almost naked of children; but with the stony, shady riversides of the Seine still trodden by a small parade of lovers, arm in arm—the perpetual fiancés of France, embracing and murmuring, content in their quasi solitude.

The widow of Georges Rouault has donated to the French state nearly all the works left unfinished at his death, some lacking only his signature to look completed—an edifying collection, to be added to the Rouaults already owned by the Musée National d'Art Moderne, and thereby make a nucleus for entire Rouault rooms in the new Musée du Xème Siècle, inspired by Minister of Culture André Malraux, which will be constructed, it is now announced, at the Rond Point de la Défense from a design by Le Corbusier. This will be the first great post-war museum in Paris. M. Malraux's other Ministerial triumph this week is the announcement that his anachronistic project for a painting by the fantasist Russian modern Marc Chagall to be affixed as a new false ceiling over operagoers' heads in the Paris Opéra will soon be under way. Envisioned by the Minister two years ago, the idea aroused such widespread shock that some leading British architects even signed a petition against it, since the Opéra is a unique example of Second Empire theatrical architecture—contestable, perhaps, but homogeneous. The present, original ceiling, by a *pompier* artist named Lenepveu, who also contributed, without distinction, to the decoration of the Panthéon, is so pleasantly innocuous that its subject matter always escapes one. What operagoers fear from Chagall is something characteristically recognizable—a green-faced violinist, perhaps, or donkeys floating in the air.

September 4

Modern French art has proved a remarkable preservative for its three original creators, that historic trio of elderly painters who, when young, fathered it in a triple paternal relationship—Matisse, who engendered the high-colored Fauves in 1905 and lived to be almost eighty-five; Braque, who created the first pure Cubist canvas, "Les Maisons à L'Estaque," in 1908 and died only last week at eighty-one; and Picasso, a few months Braque's senior, who begot proto-Cubism in "Les Demoiselles d'Avignon" in 1907 and still flourishes down in the Midi with mythological protean vigor, the one who has outlived everyone and everything, including his own euphoric painting epochs. Of the three painters, Braque was regarded as the supreme master of composition. In the purely French qualities of taste, measure, and refinement in his still-lifes, so perfectly posed as if to rest in changeless balance through time, he is ranked as the most essentially Gallic artist since Chardin, who died in 1779. By a sort of mimetism, Braque grew to resemble his own pictures, with his paint-white hair and the reminiscent cubic outline in the angular pattern of his handsome face. He was the son and grandson of housepainters and paperhangers in Argenteuil, a Seine river town near Paris, and was fortunately trained in the family trade. Expert in sign painting, he introduced the alphabet into his art like an abrupt intrusion of reality—as in the word "BAL" in his famous "The Portuguese"—and made it almost as distinguishing a mark of early Cubism as the cubes themselves. Accustomed to wallpaper and paste, his artisan hands were expert in the fabricating of those astonishing, impractical Cubist collages of paper pasteups that he and Picasso delighted in.

It was in 1907 that Cézanne's didactic maxim was posthumously published in Paris: "You must see in nature the cylinder, the sphere, and the cone," without mentioning the fourth geometric shape, which led to Cubism. That year, more than half a hundred Cézannes had just been shown in the Paris memorial exhibition that had finally given that dour painter a living public reputation, like an obituary. In the spring of 1908, Braque, who was by then a belated, dissatisfied Fauve, wrestling with wild, bright colors, went to the village of L'Estaque, on the Bay of Marseille, where Cézanne himself had worked, and, in a burst of extended comprehension, painted the revolutionary "Houses at L'Estaque"—brownish cubes floating in levitation among conical green trees. Back in Paris, Picasso was also approaching Cézanne's geometries with "Landscape with Figures," the human beings in it belonging to a new, cylindrical race, embracing in a conical glade. Unbeknownst to each other, the two young artists, both aged twenty-six, were moving toward something that was in the air but that only they were recording. It was later defined by scientists as the "space-time continuum," of which neither of the painters had even heard, nor did they ever define it aesthetically to outsiders, at least during the six years that they subsequently worked and talked together daily in Montmartre. "We said things that nobody else would understand any more and that will end with us,"

Braque declared much later of their conversations while they were painting, under various titles, what looked like Euclidean cubes and sectioned spheres against a background cascade of other cubes, producing an effect of immediate distance and the surprise of space. Among the Cubist masterpieces that Braque painted in 1910—all in museums or great private collections today—were "Still-Life with Guitar," "Woman with a Mandolin," and "Still-Life with Piano (with the keys already like nudes descending a staircase). By 1911, every avant-garde artist in Paris was a Cubist of sorts, to Braque's disgust. By 1912, Cubism had spread around Europe. Braque's works were shown in Munich with the Blue Riders. This analytical Cubism, so hermetic that it was no longer legible even to initiates, was followed by synthetic Cubism, which looked like plane geometry in bright colors—a flat, seductive, decorative patterning of commonplace objects, such as cigarette boxes or rum bottles, that even a child (an artist's child, at any rate) could have identified. Heroic, austere Cubism in space, as Braque and Picasso had practiced it, was already defunct when the war started in 1914, and their intimacy and friendship also failed to survive, perhaps because the epoch itself had perished. Doctrinarily, none of it had lived long. The Fauves—all except Matisse, who had invented them—had ceased being the Wild Beasts of bright colors. Cubism, in its inexplicable mystery and popularity, lasted in its opening unities only a few years. But it lasted long enough to start the death of representational painting, and led to the disappearance of the human being in the abstract painting that finally followed and that both Braque and Picasso, with no notion of guilt, thought pointless as painting and not art in any way. Synthetic Cubism, however, remained in general the personal style of Braque's future art life, in which his *natures-mortes* gave him worldwide fame. Whereas Picasso grew very rich, being protean as time went on, Braque, in his fidelities, became only very well-to-do. When Picasso dies, he will leave a fortune in his enormous collections of his own works. Braque at his death owned few Braques, and not even of the best.

Tuesday night, a spectacular national honor was vouchsafed Braque before the portico in the cleaned, gleaming-white Cour Carrée of the Louvre, opposite St.-Germain-l'Auxerrois, the church of the former kings, which tolled its bells as his bier was placed on a catafalque covered by the tricolor. He had been carried into the Cour Carrée by selected guards from the state museums, flanked by soldiers with torchlights. The Garde Républicaine band played Beethoven's "Funeral March." A silent crowd of many thousands stood assembled in the rain in a touching tribute of adieu. Minister Malraux had created this dramatic tableau of farewell to the painter he so greatly admired as a tribute to him and also to France. In his strange, brief eulogy, the Minister said, "Never before has a modern country rendered such an homage to one of its dead painters. The history of painting has been a long history of disdain, misery, and despair." In Braque's magisterial career and this governmental tribute to it, Malraux said, "the impoverished obsequies of Modigliani and the sinister burial of van Gogh"—one a penurious victim of tuberculosis and the other a suicide—"are revenged."

September 18

President de Gaulle, in his late-July press conference, declared, "France is advancing in great strides along the road to prosperity, and thus to power." "And thus to inflation" was the immediate corollary he should have drawn, to judge by the present hasty efforts of his Cabinet to halt France's dangerously high prices, up sixteen per cent over the past three years, and six per cent in the current twelvemonth. Paris has now been finally appreciated by its own government as the most costly capital in Europe. Because the French people are gourmets and skeptics, what first caught their national eye in the elaborate Gaullist price-stabilization project announced last week was the government promise to hold down the price of beefsteak; their immediate second reaction was to doubt that it could or would be done. To the free-spending French of today, the daily beefsteak has become what Henri IV had in mind when he optimistically talked of a chicken in the pot every Sunday. Other popular general promises are to cancel the rent raise scheduled for January and to punish the scandalous speculators in housing-project real estate by heavily taxing their capital gains. The Communists regard the whole stabilization program as *vernis*, or varnish, meaning eyewash—a cover for the major inflationary cause, which to them is de Gaulle's astronomically costly *force de frappe,* resented not only by Washington but by Moscow (and by a great many pro-NATO French, too) and loyally excoriated by the Communist Party here. The Socialists have called the government inflation-control plan a *"chef d'œvre de duplicité."* *Le Figaro,* the breakfast newspaper of the well-heeled bourgeoisie, merely called it "coherent but timid." This complex plan to check some people's spending and other people's profits from manufacturing what those purchasing people will buy cannot hope to succeed unless it is backed by the majority of the French population. It is taken for granted that no estimate of the plan's possibility of success can even be approximated until de Gaulle appeals to the nation on television, in what may once again be one of the most important addresses of his Presdential career, demanding all his authority, leadership, and powers of persuasion.

October 1

The *nouveau-roman* group has received special notice this year. "La Jalousie," by Alain Robbe-Grillet, which sold about seven hundred copies when printed six years ago by Les Editions de Minuit, is now coming out in a French paperback edition of fifty thousand copies. Mme. Nathalie Sarraute has just published a new book, "Les Fruits d'Or," and it is one of the most interesting that the group has ever produced. She and Mme. Marguerite Duras, the other woman member, seem more gifted at writing than at theorizing about writing, like the new novelist men. The Sarraute book is disconcerting, odd, absorbing. It is a novel about a novel—a novel just published and called "Les Fruits d'Or" ("The Golden Fruits")—and it tells how a coterie of literary French react to this new book, showing as much social intimacy, admiration, spite, flattery, and dislike as if it were a new personality, a person, a lover they wished to cosset or a bore they wanted to shed. The novel becomes a symbolic entity because of the way it is dealt with. Soon it is treated like a work of art—is it one, isn't it one, why or why not? This conception of it allows all the talkers to open up fully, for now they can air what they think about *l'art, la beauté.* All the tinsel, idiocy, or impressive good sense inside their minds comes out in declaration or argument; they talk louder and try to shout one another down, for each person has his or her own truth and tries to destroy the opponent. The struggle is by turns comic, snobbish, shameful, and inspiring. Mme. Sarraute (who was born in Russia) has written with such balance, subtlety, and modulation, such high rushes of mind, that the reader does not tire but floats upon all the words. In the end, the new novel is decided to be *not* a work of art, and is dropped by all except one familiar—one man who thinks that someday, in some way, he can revive it and make it a success once more, for taste is a form of time. This is the end of the book. Its inner technique descends from Virginia Woolf's "Mrs. Dalloway," Mme. Sarraute hopes, "Mrs. Dalloway" being one of her continued admirations. The book is hortly to be published in New York. It must have been very difficult to translate, but Mrs. Maria Jolas (who had much to do with the publication of Joyce's "Finnegans Wake") has put it into English of such verisimilitude that it seems merely orchestrated in another key.

October 23

Edith Piaf died at seven o'clock in the morning in Paris, and a few hours later on that same recent Friday her friend Jean Cocteau, in his nearby country house at Milly-la-Forêt, suffered a final heart attack, provoked by the news of her demise, and himself died at one o'clock. Yet at noon his incisive, familiar voice, already quasi-posthumous, was heard on the national radio among the hastily collected *hommages* to Piaf's memory, saying, "She died as if consumed by the fire of her fame." This epitaph for her had, in fact, been premature, originally prepared and recorded by him a few months earlier, when, as had been frequent lately, she seemed to be perishing and was given up as lost. But on this precise Friday, by a melodramatic coincidence, her intended epitaph suddenly became apropos for them both, in a confusion of mortal destinies.

In their opposite places in the French entertainment world, Piaf and Cocteau were known to millions, and in many cases to the same millions, both being legendary *monstres sacrés* for which the French public, from its intellectual summit to its sentimental depths, tends to share an appetite. In her last Paris triumph at the Olympia Music Hall, all that was left of Piaf was all that had ever counted—her immense, infallible voice, which rose to the roof, carrying its enormous, authentic outcry of banal phrases of anguish over lost loves, and poignant despair of happiness that would never arrive. Ravaged, ill, bundled in her customary modest black dress, her pallid little moon face set in its sad, nocturnal smile, she tottered—her thin legs supporting her in faltering obedience to her courage—across the stage to the microphone, to which she clung, and then her voice, that great remnant of her life, burst forth. Her rhythm was always like a beating pulse; she was the unassailable artist, raucously singing her *véridique* songs of truth about her own ancient class, the Paris poor. The misery of her youth became her repertoire. She was born on a sidewalk in Belleville, with two policemen acting as midwives; was brought up by her grandmother, who ran a brothel; and first sang on Montmartre street corners for sous. Highest-paid of the music-hall *vedettes,* and a tremendous favorite among record fans, she always remained poor through squandering her earnings vainly on her young men and hangers-on. The only love and fidelity she aroused were in the hearts of the vast French public—*la foule,* which retained its passion for her over the years, packing into the music halls to listen to her in delight, playing her records in lonely, shabby rooms: "Milord," "Le Légionnaire," "L'Accordéoniste," and the recent chef-d'oeuvre of them all, "Je Ne Regrette Rien," with its bold refrain, "Farewell to love with its tremolo, I start again at zero." Lately, she married a handsome Greek hairdresser, Théo Sarapo, young enough to be her son, now also turned into a singer. At her burial, in Père-Lachaise, forty thousand of *La Foule Parisienne* assembled in loyalty and curiosity to see her to her grave. On the weekend of her death, her records were totally sold out in Paris—three hundred thousand of them. As Cocteau said in that Friday epitaph, "Her great voice will not be lost."

In his elegant fashion, Cocteau was in a way the *agent provocateur* of that phenomenal revolutionary epoch of the creative mind just after the First World War, of which Paris was the geographic brain center, and which became the first—and still remains the only—style period of this century. He was a Protean, even a Procrustean, multiple-talented Frenchman, every so often cutting off the basic feet of one gift on which he was progressing toward more serious perfection so as to increase the headroom for another, which he then swelled to successful bursting with his genius for alternation, reinterpretation, and dimensional change. Essentially, he was always the same, dually composed of the brilliant lightning of his vision, in which he always clearly saw himself, and the minor thunder of his printed word, most resonant when it echoed only what he had felt in his own entity. As a precocious small boy, illicitly leaning in the early night from a window in his widowed mother's fashionable apartment near the old Théâtre du Vaudeville, he gazed down on the delicate, well-bred little mules in the private equipage that was fancifully used by and waiting for the fabulous actress Réjane, and his addiction to the theatre world began. Always in advance of the avant-garde and (though no musician himself) early tired of Debussy, Cocteau, with his infallible instinct for timing, was the organizer of Les Six, who as composers created a new, but less important, non-Debussyesque music, more useful for the Paris ballets that followed the Russian supremacies of Stravinsky and Diaghilev—to whom, in a historic liaison, he introduced Picasso as the dominant new discordant note in Paris painting. From his educated, legitimate inheritance of the Greek classics Cocteau created his own romantic bastardy, bringing the classical Greek personages down in impossible haste to the procreation of his stimulating personal mythology, which he used to people his plays and films, with their old truths and mysteries of misconduct—the Sphinx, Oedipus Rex, Antigone, Orpheus—and from there farther down to earth, among mere human beings and the old filial and parental errors of "Les Enfants Terribles" and "Les Parents Terribles." He began the fashion for the absurd in his creation of "Les Mariés de la Tour Eiffel," and enlarged the modern appreciation of death in the suicide-by-hanging of the *premier danseur* in the ballet "Le Jeune Homme et la Mort." The vogue for violence was early manifested in his film "Le Sang d'un Poète," disapproved of by the Church. Films were pristine material then. On them he left his elegant thumbprint. He illustrated some twenty-five books, mostly his own; his familiar profile drawing of a round-eyed, short-lipped Orpheus was so famous that in Cocteau's honor the French state engraved it on a twenty-centime stamp, as if it were his own portrait. His most serious volume of poetry, "Plain-Chant," should long outlive him. "Thomas l'Imposteur" was his earliest, most important novel; "La Difficulté d'Etre" was his richest autobiographical report. And in the realm of phantasmagorial reality in that strange 1930 Paris literary epoch of drug addiction, the text and frighteningly informative physiological drawings of his book "Opium," created in a disintoxication clinic where he did not sleep for twelve days, are of a rare literary as well as psychiatric importance. In this book he defined opium as "the sole vegetal substance which communicates the vegetal state; through opium we obtain an idea of that other speed known to plants." As a painter, he decorated two country Catholic chapels, one of them St. Blaise-des-Simples, a twelfth-century lepers' retreat in his village of

Milly-la-Forêt, on whose walls he painted frescoes of herbs and simples formerly deemed helpful to the sick—the polydore fern and *absinthe artemisia.* He was buried in its churchyard. His beautiful hands, which he displayed in narcissistic admiration during life by wearing his cuffs rolled back to uncover the artistry of flesh and bone, were on view for the last time in the pictures taken of him on his deathbed.

November 26

Never before in our time have the French been so unified in a sympathetic public emotion as in their grief and shock at the assassination of President Kennedy. Not a sentimental people, but experienced in tragedy, they seemed to feel that his death summed up, like a legend, all that must be mourned—the young leader of his country vilely, treacherously cut down in his vigor and prime; his wife, in her beauty, a widow; his small children fatherless, the melodrama of his dying a reproach to his own land; and what he left behind him still undone in his pursuit of his hopes a loss to all men of good will. These were the causes of their grief, which Parisians talked about, often with visible tears. His death loosed emotions here, and also suddenly clarified appreciations. About these the humbler French were especially explicit, giving the reasons for their faith in him. They said he had worked without rest for peace between West and East. A rich man of the richest nation, he had had compassion for the poorer nations, had shown wiser statesmanship and more diplomatic patience than men nearly twice his age, and, above all, had fought as much as he was allowed to for civil rights, for equality between men regardless of color. And he had had faith in God. This last, coupled with the French knowledge that he was our first Catholic President, made him seem more comprehensible to them, more touched by grace, than Franklin Roosevelt, heretofore their favorite President of this transatlantic epoch.

Now, less than a week after the fatal event, already a certain realistic alarm amplifies the French grief. Whatever President de Gaulle's cumulatively successful projects for the self-domination of France and for its domination of Europe, Kennedy, as President of the United States, was the actual leader of the Western alliance, which includes France within Europe, and in his death the French feel temporarily uncovered, almost denuded. Parisians, in their cynical acuity, saw the irony of de Gaulle's attending President Kennedy's funeral in Washington after having earlier discouraged his proposal of a visit here. But to the majority of the French, de Gaulle's presence at Arlington was no mere formality between heads of state, living and deceased. It was a natural family gesture of sympathy between France and the United States, which, had he failed to make it, would have shocked his people as an omission of respect for the admired dead President. The only lessening here in the tragedy of Kennedy's assassination is that it has helped restore the feeling of Franco-American amity.

His noblest epitaph, from the pen of the French poet and former diplomat St.-John Perse, has just been published by *Le Monde,* entitled "Grandeur de Kennedy." It opens thus: "History created no myth. Face to face, he was a man simple, close, and warm, prompt to the activity of each day." It continues, in part, "He was the athlete racing toward his meetings with destiny. He fought always with his weapons unhidden, and in his meeting with death his face was uncovered. Upon events he imprinted a mark of progress that was his own and that leaves us following his path. At the service of a great people in love with liberty, he was a defender of all rights and all freedoms. No one was more the enemy of abstractions or more carried by instinct to the heart of things. He had the clear, direct gaze of those young chiefs formed for friendship with mankind. When fate lifts so high the burst of its lightning, the drama [of death] becomes universal, and the affliction of one nation becomes that of all."

1964

February 11

At his recent Palais de l'Elysée press conference, President de Gaulle appeared for the first time to have aged, at one moment resting his face on his hand in the way a reposing old eagle leans his beak on his neck plumage. During the first forty minutes of his monologue, or until he started talking of his recognition of Communist China, the foreign journalists present were politely restive and bored as he concentrated on his ideas of his supreme Presidential and Constitutional powers over France—ideas that, because of their extreme candor, were of considerable interest to the French listeners, at least. What he was saying was, in reality, directed at specifically one Frenchman, who was not even present. This single Frenchman was M. Gaston Defferre, Socialist mayor of Marseille and sole announced candidate so far for the Presidential elections to be held at the end of next year, probably against the unbeatable de Gaulle, who will then be seventy-five. In all that de Gaulle said for Defferre's information on how republican France has to be handled, "rarely," commented *Le Monde* the next day in a stern, scandalized editorial, "has the theory of absolute power been revealed more complacently, clearly, or rigorously." The paper continued, "It is a good thing to vituperate against the impotence of government by parliament and to recall its miseries, but it is no less necessary to denounce the dangers of reactionary excess. If one admits that everything in a country may depend upon one man only, a more or less totalitarian dictatorship is already present in germ."

On this past Saturday, Defferre opened his Presidential campaign by stumping in Bordeaux, a Gaullist stronghold. There, according to reports, he seemed both successful and unusual as a candidate. (He had already stated, with a sportsmanship rare in French political circles, that to be defeated by de Gaulle would be "no dishonor.") In his afternoon speeches, he intelligently chose to talk to college students from the Institut des Sciences Politiques and to workers from the Socialist labor union and the Catholic labor union, workers from the Communist-led C.G.T. union having refused even to come listen to him. That night, he collected an audience of nearly four thousand average citizens in a Bordeaux suburb, to whom he courageously declared that, as of that evening, they were no longer faced with the prospect of "de Gaulle or nothing." He rejected de Gaulle's notion that the Presidency should furnish the control and also the source of all power, which Defferre said sounded like "an absolute monarchy," as if French history had not already shown what that led to. His keynote policy was against "false French grandeur," and he further declared that he was against the French *force de frappe,* preferring a common nuclear policy for a powerful, economically and politically united Europe, including Britain and Scandinavia, which would be of a Socialist character and would favor the common interests of all countries rather than the special interests of a couple of powerful states. As for his far-off electoral struggle against de Gaulle, Defferre optimistically referred to that ancient combat between David and Goliath.

Thus, the end of 1965 will furnish France with the first direct popular election of a President since that of Louis Napoleon in 1848, who for good measure was elected Emperor four years later. Defferre is modelling his campaign on that of Senator John Kennedy in 1959–60 and, in imitation of Kennedy, has started nearly two years before the election date, in a country where campaigns for senators and deputies—about all the French are accustomed to elect—rarely last more than a fortnight. Again in contrast to usual French political practice, Defferre has actually got his Socialist Party (of which he is, after all, the Parliamentary whip) to agree that he, as the candidate, will set the campaign policy, and he has already turned thumbs down on the diehard Socialists' fatuous hopes of a return to the fatal system of government run by Parliament. Like most anti-Gaullists, he was flatly opposed in 1962 to de Gaulle's Presidential innovations. Now, like most of the more modern-minded middle-aged French, who have had to learn their modernism quickly, he believes the

Presidential system is here to stay in France. If he were to win, Defferre says, he would not even change the number of the republic; it would continue with the title that President de Gaulle gave it—*la Cinquième République Française.*

"Parlez-Vous Franglais?" is the title of an entertaining, if repetitious, book that is causing a lot of chatter here. Its author is M. René Etiemble, professor of comparative literature at the Sorbonne, until now best known for his studies of the poet Rimbaud but lately launched against the corruption of the French language by its postwar inclusion of Americanisms, which produces a bastard transatlantic tongue that he calls Franglais. "Language is the blood of a nation," he solemnly declares. "Since the Liberation, our blood has been much diluted. The vocabulary of the young generation that will be twenty years old in 1972 is already one-fourth composed of American words. At twenty, these young people will not be able to read Molière, let alone Marcel Proust." His book cites hundreds of Franglais words or phrases used in every walk of French life today, beginning with his opening chapter about *les babys* and the *coin de teens,* or teen-agers' corner, from which it moves easily to bar drinkers' requests for *"un baby Scotch sur les rocks." Le sport,* which the French took up late compared to the British and Yanquis, as Etiemble calls us, is rife with Franglais, such as *"les trottings"* at the race track, *"un crack,"* for a topflight jockey, *"un coacheur," "le karting,"* and *"les supportères"* of the home Rugby team. There is also the old Franglais phrase *"faire du footing,"* which means merely to take a walk (on your foots, naturally). Big business in Paris now features a weekly *réunion de briefing* in office buildings *de grand standing.* A millionaire executive's yacht is called that but is pronounced to rhyme with "watch." Hollywood camera terms are used even by teen-age movie fans here, such as *"un travelling,"* and so on. In the intermission between films in the Champs-Elysées movie houses, the girl ushers now sell a nut candy they loudly offer as *"noots,"* which always breaks up us Yanquis present. What most enrages Etiemble, probably a refined, slow-eating gourmet, is the old *bistros* modernized under neon signs as *"le snack," "le quick,"* even *"le queek"* or *"le self,"* which means a cafeteria, and even *"le self des selfs,"* on the Boulevard des Capucines, which means nothing on earth. As a philogist, he seems not to note that French language and cooking lack the word and the celerity for out snack-bar fare, which young Paris office workers immediately developed an appetite for at Le Drugstore, on the Champs. Etiemble concludes dispiritedly that the future of the French language is English. Alas, we Anglo-Saxons don't export it; it is the French who import it. This is an aspect of the Franglais problem that the Professor neglects to mention.

An exhibition of sixty-eight new Picasso pictures, from 1962 and 1963, has just opened at the Leiris Gallery, under the aegis of M. Daniel Kahnweiler, his early art merchant, who knew him just before his Cubism began, in 1908. There are still traces of it half a century later in what the painter has painted in his early eighties. For the past ten years, perhaps, he has been painting in a consistently convulsive manner, with traces of everything consequential indicated, if not expressed. What astonishes in these new pictures is their aesthetic energy, which the viewer can see with his eyes just as surely as he can hear with his ears, at a concert, the musical energy blaring forth from Verdi's trumpet scoring—a thrilling creative vigor being part of the sights offered by the Spaniard just as it is part of the brassy sounds supplied by the Italian. Picasso's special repetitive theme this time, of which he shows thirteen versions, is that of the painter and his model. Some Paris critics have fallen to ruminating about monotony in this 1964 Picasso exhibition. The truth probably is that Picasso is an old re-creative virtuoso, and his critics are now only wearied, practiced onlookers where he is concerned.

February 26

When President Antonio Segni, of Italy, was here on his recent official visit to General de Gaulle, it was planned that the visitor, who is a distinguished former professor, would visit the Sorbonne and be welcomed by its students. They, on their side, planned to greet him with the cry "Monsieur le Président, behold our university in ruins!"—referring to the dilapidation and crowded classrooms of their august institution of learning, which was founded in 1256. So the university authorities ordered a one-day lockout, and last Friday, when Segni called, all the Sorbonne's buildings were closed against its scholars and the students retaliated with a one-day strike—their customary retort to official pressure in their running quarrel with the state and municipal authorities. Yesterday there was a monster protest meeting of students in the court of the Sorbonne, and today the government's Conseil des Ministres took up consideration of the long-awaited plan for a general bureaucratic overhauling of French national education proposed by the unpopular Ministre de l'Education, Christian Fouchet. It certainly seems inappropriate that in a country haughtily proud of its cultural superiorities and tradition the normal relation under the Fifth Republic between the scholars and the state whose colleges furnish the means of culture should be that of a perpetual cat-and-dog fight. Students old enough to vote, and their parents, have been waiting to see whether Gaston Defferre, the Socialist candidate for President in next year's elections, would include in his platform, criticizing the de Gaulle policies, a plank that would deal with the Sorbonne scandal and offer a way out of the impasse.

As a matter of fact, Defferre, except through provincial newspaper reports of his speeches in the southern country towns where he has been stumping, is not having an easy time letting the French people know what any plank in his platform will be like. In Trèbes near Carcassonne, on Sunday, at a Socialist Party luncheon attended by twelve hundred members, he declared that the government had given orders to both state-run radio and television to boycott his meetings and speeches. Spotting two TV cameras aimed at him, as if in rebuttal, he called out, "They're not French! One is for a German news program, the other is English!" He also made a speech at Narbonne, but citizens who were listening to the radio never found out what he said in it, for that was not put on the air, only the fact that he spoke at all being considered unavoidably newsworthy. Ever since French radio and, especially, television began expanding in political and newscasting importance, as a result of the brilliant use that de Gaulle made of them, French citizens have become increasingly conscious that these not only constitute the most powerful government monopoly of mass media but also are controlled—meaning censored. The sudden national political existence of Defferre as a candidate who is criticizing the President's policies in his speeches has subjected the whole situation to heated controversy. It was noted that when he announced his candidacy, a month ago, and said that the French, aside from the citizens of Marseille, where he is mayor, did not even know what he looked like, the state television showed only a most fleeting glimpse of his face, as a national introduction, and finished its camera shots of him from behind his back. Criticism of the party uses that the state television and radio are put to here is, naturally, not new. When Guy Mollet, a Socialist, was Premier, during the Fourth Republic, the tomatoes thrown at him in Algiers by the angry Algerian French were reported on fully in the French newspapers and pictured there, too, but certainly were not featured on radio or TV. The heated discussion right now about unfair radio and TV coverage for Defferre is probably overhasty, especially since it has already been announced that the 1965 campaign will officially last only two weeks, with each candidate being given two hours' radio and television time, to be spread over the fourteen days. De Gaulle, of course, can always procure the lion's share by, in addition, giving one of his twenty-minute TV talks to the nation, as the incumbent President.

Actually, the government broadcasting monopoly is by no means ironclad as far as listening is concerned. It has lively competition from Radio Luxembourg; from Europe No. 1, whose station is in the Saar; and from Radio Monte Carlo—not to speak of the B.B.C., which ever since the war days has been the French favorite for honest news. But there is no kind of TV here except the Fifth Republic's TV. There are now five million television sets in use here, and the number is constantly increasing, which means that more than half the French can be reached visually on the screen—and influenced. However, even this absolute monopoly, sinister as it is in principle in a republic, has its comic side here in France. Government officials privately insist that the state control of television only evens things up for the regime; because the majority of the newspapers are against the regime, monopoly TV is the only chance it has of giving itself a fair show.

March 11

Local diplomatic circles are now saying, in their professional undertone, that Franco-American relations have deteriorated to a point where little is left of them except formal recognition. There is supposedly a governmental anti-American atmosphere that is à la mode in Paris. As yet, this erosion on the government level of the transatlantic national friendship has in no perceptible fashion influenced the nationals themselves here—the vast population of

Parisians and the small colony of resident Americans, who continue to get along in their relative separateness exactly the same way and exactly as well as they did before General de Gaulle recognized Peking, which is where it all started. No sooner had the Chinese delegation settled down temporarily on the second floor of the Hôtel Continental, across from the Jardin des Tuileries, than de Gaulle gave a particularly cordial welcome to M. Nicolai Podgorny, Secretary of the Soviet Presidium, as head of a Soviet delegation that enjoyed a gala visit in Paris, with Podgorny recalling that President de Gaulle had a standing invitation to a gala visit in Moscow. Then, this last weekend, millions of French TV watchers saw and heard Dr. Castro, of Cuba, where de Gaulle is to visit in a few days, give a televised, translated interview in which he was warm indeed in his admiring references to the Fifth Republic's chief. "There are several things in General de Gaulle's foreign policy with which I am in sincere sympathy," Dr. Castro said, and and enumerated them. First of all was the French leader's independence of attitude in regard to the United States, then his opinions on how to settle things in Southeast Asia, and finally, of course, his decision to send a French Ambassador to Peking. For all three of these anti-American compliments to be programmed in French on the French-government-controlled TV was like a triple massage of salt in Washington's wounds.

In confused and worrisome periods like this last fortnight, with Ambassador Charles Bohlen known to be back home talking over the Franco-American rift with the State Department, Parisians have been extra diligent in their perusal of the papers to find out what is going on in national policy matters—any papers, even the British press when it contains special pieces on French affairs. This Tuesday evening's *Monde* hastened to print in translation a Monday-morning editorial from the London *Times* entitled "De Gaulle Leans Left." With dry humour, the *Times* said, "Suddenly President de Gaulle seems to be the only active revolutionary in Europe. He has just completed a remarkable double in recognizing Peking and acting as host to a high-powered delegation from Moscow. From many a foreign spokesman of the militant Left come tributes both to the historical role of France and to the President's reinterpretation of it. At first sight it is an odd part to be played by a man whose own domestic opponents caricature him as the reincarnation of Louis XIV."

Also on Tuesday, there was a bold, analytical anti-Gaullist editorial in *Figaro* by M. Raymond Aron, Sorbonne professor of sociology, a graduate in philosophy, and a multilingual intellectual of the type that the French admiringly call *un mandarin,* who was formerly a Gaullist himself and is today the French commentator on national and world affairs best known in Paris and also in Washington. Aron titled his editorial, as if in inquiry, "The Infantile Malady of Old Nationalism?" and began his analyzing with several other pertinent questions. "Are the French in 1964," he asked, "more nationalist or less nationalist than in 1958?"—when de Gaulle came to power. "Are they more or less favorable to a united Europe? More or less hostile to the Atlantic Pact and the United States? Personally, I would not presume to decide, though in political milieux the opportunists seem convinced that to tag themselves as anti-American is the best way to court the Prince"—as de Gaulle is semi-ironically called. "What characterizes French nationalism today is less the state of mind of the French than the diplomacy of the Republic's President," he went on. "At every turn, it keeps Washington at a distance. Faced with the Soviet's exigencies, it seems ready, despite everything, to renew the traditional old game on the planetary system of drawing near first to one body, then to another—the only principle being never to be definitely tied. [In this] neo-nationalism of *la France éternelle,* she is on the hunt for a first-ranking role on the world scene [and appears as often as possible as] a lone rider, so as to wipe out any appearance of a connection with the Atlantic Pact." Aron added, "Thus, the world sees what she does as an expression and a justification of the revolt against 'American imperialism.' That France is concerned about herself, everywhere and always, is too well known to need repeating. The question the bystander asks himself is: What good do some of these decisions do for France, aside from their undeniable merit in irritating both allies and adversaries at the same time? What will France gain by making the People's China a member of the United Nations and exasperating an important segment of American opinion—aside, of course, from the glory derived from showing her indifference to the susceptibilities and preferences of an ally whose major error is that it is in possession of the reality of power?" Aron's last, highly important, question was to ask, "How does Gaullist France see the international future? In terms inherited from the Common Market and the Atlantic Pact? Or, rather, in terms that suggest the exaltations of nationalist grandeur and the subtleties of a diplomacy born in the Italian cities of the Renaissance? If France herself shows today certain symptoms of the infantile sickness that is nationalism, what country tomorrow will escape contagion?"

Ideas on the future of France, but of a very different sort, were delivered last week (and much commented on) by another highly esteemed intellectual, who is professor of law and economics at the Sorbonne and director of social research at the *Sciences Po'*—M. Maurice Duverger, ordinarily a front-page political analyst for *Le Monde*. What he had to say on Franco-American affairs was printed as an interview in the stimulating leftist weekly *L'Express,* which is anti-government and, indeed, anti-everything except criticism and vigorous information. The essence of what Duverger passionately declared was "It must be said, it must be written! There is only one immediate danger for Europe, and that is the American civilization. There will be no Stalinism or Communism in France. They are scarecrows that frighten only sparrows now. Today, all that belongs to the past. On the other hand, the pressure of American society, the domination of the American economy—all that is very dangerous. Why do you make such a face?"

To this, the *Express* reporter replied, "Because anti-Americanism gets on my nerves. Everybody in France has become anti-American. De Gaulle, Etiemble and his 'Parlez-vous Franglais?,' all that talk about 'Oh, my dear, China!,' and 'The Americans, those barbarians!' *Zut!*"

"Granted," Duverger replied. "Nothing is stupider than stylish anti-Americanism. But at the base of it all there is, just the same, a real question. America is a different society from ours. It was built by pioneers who for their cultural baggage had the Bible and a sense of adventure. With these two elements, they succeeded in making a body of men for whom money is the essential criterion—the basis of their system of values. In a country like France, the employee who reads your gas meter possesses a scale of aristocratic values. He can distinguish perfectly among a *nouveau riche,* an intelligent man, and a poet. Whether he knows it or not, the cultural ensemble that is at the bottom of his attitudes is shaped by an accumulation of history different from that of the Americans. I think that this element will help us to resist the American pressure. But don't forget one thing: America is evolving. As Russia is liberalizing itself, so is the United States civilizing itself."

"It is doing it pretty rapidly," the recalcitrant reporter interjected.

"In some sectors," Duverger admitted. "But it will probably take as long for America in its entirety to reject a system of values based on money and gadgets as it will take Russia to reject a political system based on dictatorship. For us, the essential is to escape both of them. Our luck, as Europeans, is that we are behind the times—in planned consumption in relation to the United States and in planned socialization in relation to the U.S.S.R. For us, the problem is to arrive at the abundant society when the transitory American phase is over and at socialization when the transitory phase of the proletarian dictatorship is finished. The second seems sure. The first is not sure at all. Quite the contrary. That is why America is for us the more dangerous."

Painful as many of Professor Duverger's conclusions are for many Americans, he has academically touched on basic, alarming truths for many of the French, who, even in their awareness, seem unable to do anything about them except complain—while continuing their American way of life *à la française*.

March 23

"La Damnation de Faust," which Hector Berlioz completed in the middle eighteen-forties, a weak period in French music, was indecisively described by him as *"une symphonie avec programme plastique."* For the first time in musical history, it was recently presented at the Opéra in precisely this form by the *metteur en scène* and choreographer Maurice Béjart, and the fashionable first-night audience and most critics took it to be an extraordinary mixture of splendid and shocking ballet plasticities, of scenic brilliance and muscular bad taste, of aesthetic beauty and willful uglinesses. It was a plenitude that led to noisy dissension, with outcries, catcalls, and insults being freely launched at the performers and exchanged between the spectators who applauded what they saw as something astonishingly creative, at its best, and those who booed it as merely scandalous. At the second performance, the quarrel was still going on, but more mildly. Those who disapproved shouted, *"C'est indécent!"* The admirers then shouted back, *"A la tombe les ancêtres!"*—the anathema against elderly conservatism lately popularized by the younger French and meaning, very roughly, "Off to the cemetery with all the old fogies!" So little that is remarkable ever takes place at the Opéra, aside from the amazing mediocrity of the singing, that to be credited with a scandal gave its reputation an invaluable fillip and "Damnation" a run at the box office.

What Béjart had ambitiously aimed at was to have a ballet illustrate the temptations, visions, and earthly events connecting Faust and Marguerite, which Berlioz' singers merely refer to in solos. To give his dancers room, he immobilized the choruses (dressed like medieval monks) and also made Faust and Marguerite practically stationary, as if in an oratorio—the form in which the opus used to be given. In Act I, Faust was discovered hung aloft above the footlights in a cage, from which he motionlessly viewed and sang about the sad state of the universe. It must be frankly said that Béjart has publicized his career as a stage director in part by his scandalizing imagination and his eccentric fixations, such as hanging things from the ceiling, and in his choreography he has been addicted to choruses that hop about in a semi-squatting frog position. For the famous "Hungarian March," his ballet of warriors strode in like Storm Troopers, dressed in belted khaki-colored tights and wearing what looked like motor-cop helmets decorated with death's-heads. Then, in a froglike pastiche of totalitarianism, they captured Le Jeune Homme, or Youth, who symbolized freedom, in the person of Cyril Atanassoff, the Opéra's excellent new male ballet star, who has long been needed there, as has the new star ballerina Mlle. Christiane Vlassi, a slender, ephemeral vision symbolizing the soul of Marguerite. Paris had never before seen the Fifth Republic Opéra Ballet perform as a whole with such impressive, superior technique, constantly dancing far better than the Opéra singers sang. (At the end of the first performance, when the cast made its bows, the unfortunate tenor who sang Faust was booed by the gallery gods.) In the most elaborate, poetic, and aesthetic ballet number, a half-dozen delicately bodied young coryphées wearing nothing but flesh-colored tights danced as nymphs in the half-light, in seeming innocent nakedness among the mortals. This ballet was like a Botticelli scene. Then, as Faust's tastes became more earthy, equally nude-looking girls under blowzy wigs lubriciously saddled themselves around the male dancers' waists, like infernal figurines from a Hieronymus Bosch canvas depicting antics in Hell. The décors, by Germinal Casado, Béjart's assistant, were mostly stunning—especially that of the town walls, which shone metallically, as if covered by copper. And Marguerite's small Gothic house suddenly split in two to reveal a tall silver tree beside her opened bedroom, where she sang the celebrated "King of Thule" aria.

In the last act, as an added surprise, Faust and Mephisto were seen seated on a pair of metal horses en route to damnation, contrary to Goethe's drama, of which Berlioz used the madman Gérard de Nerval's French translation as his libretto. One Paris critic was worried by the possible redundancy in Béjart's scheme of duplicating the singers with dancers, so as to make the action extra clear. For most of us there was no such pleonastic danger.

When President Nicolas Grunitzky of Togo, a worldly, smartly dressed pale African of mixed German, Russian, and African blood, made his state visit here recently, and twenty-five gigantic, costly Togo national flags (yellow and dark green, with a square upper corner of revolutionary scarlet mitigated by a big single white star of hope) flapped from the Concorde mastheads, and thousands of Togo bannerets fluttered over the Champs-Elysées and on the fronts of municipal autobuses, Parisians knew, by their ignorance of who on earth the visiting distinguished head of state might be, that he must be another remnant of European colonialism now practicing *le Présidentialisme* and come to call on General de Gaulle, his top hat doffed, for a handout—another in a regular parade of them here over the past few years. (Owing to the supposed historical fact that a prewar native deputy from one of the French colonies was invited by his tribal constituents to lunch and became part of it, some cynical Parisians now refer to these African débutant democracies as *"les républiques cannibales."*) In a series of elaborately documented articles in the illustrated, influential weekly *Paris Match,* entitled "Attention! La France Dilapide Son Argent," M. Raymond Cartier, the magazine's chief editorialist, has established the economic doctrine of so-called Cartierisme, which, as a criticism of the government's spending policy, has caused heated comment and worried smaller editorials all over France. As his main thesis, Cartier points out that France has been spending more than two per cent of her gross national product on foreign aid to underdeveloped countries—over three times what the United States has spent of hers—and that France had better look first to her own underdevelopment, which he then paints in. The Sorbonne classrooms, he says, are "as crowded as the Métro;" no new Paris hospitals have been built in thirty years; "Asiatic conditions" exist in small French villages, almost forty per cent being without running water; even in big French cities, thirty-one per cent of the apartments are without private water closets; and France's telecommunications are the poorest among all the European Community nations. Cartier ironically advises the hundreds of thousands of the French who are still waiting for their phones to be installed to "go to Abidjan, where ten thousand telephone lines, without one user," have been installed with the French taxpayers' money. In real distress, he deplores the fact that a fabulous national French fortune has been splashed over the ex-colonial lands, "with an air-conditioned Versailles palace for the Ivory Coast president, and little of the aid money in general ever reaching the poor inhabitants, who are often too illiterate to read their phone numbers and, in some places, so confused about putting ballots into urns that they add their marriage licenses." These and a myriad other costly policy errors, Cartier says, restrict the French government on its home circuit to "a Malthusian economy." As for the colonial past, he says that France should not "take the moral attitude of guilt today," for, with the possible exception of ostrich plumes, "all the colonial riches were created by European capital and work." And, as he failed to add, European capital pulled in the profits.

April 7

By the energy of his imagination and the generosity of certain of his cultural practices—frequently startling to the French, such as his having lent the Venus de Milo to the Tokyo Olympics—M. André Malraux, as Gaullist Ministre d'Etat Chargé des Affaires Culturelles, is unquestionably the most discussed and most powerful figure in European culture today, and certainly the only one backed by a government. A dazzling conversationalist who never gives interviews, he recently answered a few questions put to him, over coffee, as a luncheon guest of the Anglo-American Press Association here, the answers swelling to a brilliant, important post-prandial kind of lecture—fortunately taken down in stenotype, since the scintillating speed of his thinking and speech was beyond any of us in the way of adequate note-taking. In answer to the opening question, put by the London *Observer's* Paris correspondent, asking what M. Malraux would think of a recent suggestion that Great Britain create a Ministry of Leisure (which France had already invented and temporarily enjoyed under Léon Blum and the Front Populaire back in the nineteen-thirties), he replied, in part, "This question is one of the most important concerning the human mind and thought in this half of our century. But we must first find out whether these problems of ours should come under the heading of leisure. I think they should not. For probably ten thousand years one civilization was very much like another, with the chief of state dealing basically with agriculture, the police, the army, and finance. I will repeat what I said to President Kennedy, what I have said to the French Parliament, what I have said all my life—that throughout all that time the same things dominated humanity. If a man like Rameses II of Egypt had found himself face to face with Napoleon, they would have talked about almost the same topics. One day, all that came to an end—and that day coincides with us. What changed everything? You all know what it was. The machine, naturally. This is where the essential problem arises. If the machine is victorious, it creates spare time and, consequently, what is called leisure. The machine creates objects but it also creates the multiplication of dreams. A hundred years ago, some three thousand people in Paris went to the theatre every evening. Today, if you count television, how many people in Paris see a show every night? Probably three million. You can put unimportant things on the screen. You can also put on comedy. Comedy is so important that it alone has unified the capitalist world, doing what the will to revolution did for the Communist world. In the cinema, there is, on the one hand, Chaplin and, on the other, Eisenstein. In between there is nothing with a comparably profound effect on man's sensibilities. But, comedy aside, what remains is the essential, what I have called the eternal—to put it clearly, the realm of sex and the realm of blood. Make no mistake about it, modern civilization is in the process of putting its immense resources at the service of what used to be called the Devil. There is a great domain of nocturnal darkness in man, which none of us can mistake if he looks at himself in the mirror; even the most cretinous people, out to make money by exploiting this among defenseless children, recognize it as clearly as we do. What counts against the appeal that the power of money holds for the powers of darkness?

Here is the only great problem that exists for me behind the word 'culture.' "

The Paris correspondent of the New York *Post* then said to Malraux, "Monsieur le Ministre, a united Europe is being created culturally as well as politically. How do you envision a Europe in which nationalism has been left behind?"

To this Malraux instantly and elaborately replied, "Politically, I don't think it true that nationalism is outdated. I think we are involved in an appalling historical misunderstanding. The nineteenth century considered that nationalisms were provisional hypotheses, and that the twentieth century would be internationalist. That was not meant as a joke, since even Victor Hugo thought it childish to conceive of our century as an era of nations. Nations were what the provinces formerly had been; at the end of the nineteenth century, the provinces had become nations. Thus, the nations would become Europe. Opposed to all these ideas was somebody, who was by no means a nobody, named Nietzsche, who said, 'The twentieth century will be one of national war.' Well, has the twentieth century been one of internationalism or one of national war? Has the Soviet Union become Russia? Leather-jacketed People's Commissars have become gold-braided marshals, but is that really internationalism? Has China, the most international country, become especially internationalist? Or Italy? Or Germany? In our time, the century's dream has turned out to be nothing but the drama of nations. We are in the century of nations, and our predominant problem is how to reconcile the essential reality of nations with our hope of happiness for the world—the hope for justice, which men cannot establish except on infinitely vaster structures. To sum up clearly, I think the notion of Europe is a fundamental one. But I think that to believe that an over-all Europe will be brought about by a handful of people who simply agree about it is absolutely puerile. The destiny of the world is made by hard realities, not by good intentions. Something very strange is happening in the realm of human thought. We have been initiated into the fraternal elements of the great religions. When the great religions were great, they hated each other. But the moment they found another enemy, the wars of religion ended, with reconciliations among Catholics, Protestants, and so on. For the first time in this century, there is a domain in which fraternity is possible—a domain containing all that is most mysterious and profound in our heritage in this field of thought. It is the most powerful ferment of the future Europe."

Close to five thousand people assembled at Père-Lachaise Cemetery this last Sunday morning in the sunless, wintery cold that has marked this spring to assist at the dedication of a monument to the memory of the fifty-six thousand French deportees who died at Buchenwald. Slowly paid for over the years by public subscription, the monument is a terrifying modernist statue of two skeletal men, starved close to death but still upright and sustaining each other. It is the work of Louis Bancel, a French Buchenwald prisoner who survived. On either side of the monument, for the dedication, were placed temporary urns with little smokestacks, from which wisps of smoke emerged during the services, in faithful recollection of the Buchenwald crematories. The opening dedication speech was made by the elderly Abbé Blanc, who had been the clandestine almoner at the camp. He began by saying familially, "I am addressing the mothers and fathers, if any are still alive, of those young men who died there. I am also addressing their children who may be here, and their widows." These were many, to judge by the signs of grief manifested. Most of the middle-aged Frenchmen present (they had come from all over France) were camp survivors, who reminisced together or could be heard suddenly recognizing each other—*"Tiens, c'est toi!"*—as they met for the first time since they were liberated nineteen years ago on Saturday of this week. It was a most affecting scene and ceremony for all present.

April 23

The uglification of Paris, the most famously beautiful city of relatively modern Europe, goes on apace, and more is being carefully planned. Already, one of the enormous square blocks of skyscraper office buildings on the Boulevard de Vaugirard, which forms the first part of a huge complex of modern buildings that will be composed around and will include the new Montparnasse railroad station, is, with its eighteen stories of unmitigated cement, a solid, high eyesore on the southern skyline of Paris. It disgraces the horizon as seen from the city's upper windows giving on the Tuileries Gardens, which were formerly possessed of one of the capital's most elegant perspectives. For these ignoble changes, big business and the automobile, either squatting by the curbs or crawling or racing along the boulevards, are responsible. Now there is a plan to create a double motor highway by the Seineside, from the Place de la Concorde to the Pont Sully, along the Right Bank. Already, members of the Institut and the Academies have written eloquently in protest to the helpless Ministry of Cultural Affairs. To build the motor highway demands the sacrifice of the elderly, drooping elms that for decades have leaned over the river, opulently rooted and watered beneath the heavy cobblestones of the verge—the quietest promenade for lovers or solitary thinkers in all Paris. Cutting the elms, now in their faithful spring verdure, seems, in anticipation, a murder of nature, beauty, and art.

May 5

UNESCO held a two-day philosophical colloquy a week or so ago to mark the end of L'Année Kierkegaard, the European celebration of the hundred-and-fiftieth anniversary of the birth of the famous, eccentric Danish progenitor of Christian Existentialism. With the brilliant Tuesday-night lecture on him by Jean-Paul Sartre, which marked this French writer's return to the public scene that he formerly dominated in Paris like no other academic figure; with messages (in part recorded) in German from the German Existentialists Martin Heidegger and Karl Jaspers; and with platform participation by such leaders as the Italian Enzo Paci and the French professors Jean Wahl and Mlle. Jeanne Hersch, the handsome, modernist UNESCO auditorium became, as one philosopher admiringly remarked, an Olympus for the most eminent masters of contemporary Existentialism, still regarded in Europe as the most authentic modern philosophical thought. There was a slight hitch in the Tuesday-evening proceedings when the closed doors were assailed by a troupe of young Sorbonne *philo* students without invitations, who, upon clamoring "We are philosophers, too," were admitted. UNESCO dignitaries squeezed over on their benches to make room for youth, or else the students sat on the floor in the aisles, where they had a fine view of Sartre as he read his long address, he being the star they had come to gaze at with silent respect and to hear. He read his lecture in a quick, pleasant voice, with useful, hammerlike diction. Since Kierkegaard himself had famously said, "Truth objectively fixed is paradox," Sartre felt free to pour out paradoxes and truths like champagne. He said that the title of the colloquy, "The Live Kierkegaard," was proof that he was indeed dead but still influential; that "he was a philosopher who detested philosophers"; that "he stole the language of knowledge to use against knowledge," meaning what Hegel then stood for, "a system for thought but not for existence," from which the unorthodox little Dane's original Existentialism sprang; and that Kierkegaard "refused to the serpent the right to tempt Adam—as if God had need of sin!" Speaking personally as "an atheist of the twentieth century," and drawing largely on his own theories as laid down in his "L'Etre et le Néant," Sartre also poured out, among other shining formulas, "Contemporaries understand each other without knowing each other," "The liberty in each man is the foundation of history," and "Man is the being who transforms his existence into intelligence." In appreciation of the fact that those two so different nineteenth-century rebels, Kierkegaard and Marx, were writing their tracts of social revolt in the same years, Sartre added, "These two *morts-vivants* control our existence today." Heidegger, in his message to UNESCO from Germany, said, "The development of technology has put the finishing touch on philosophy. Philosophy comes to its end in the present epoch." UNESCO will publish the Sartre lecture at the end of the year.

May 20

Paris has grown to have such endless suburbs that the Départements surrounding it, and its own Département, the Seine, are now apparently going to be split up, and given slightly new shapes and new geographical names. There will be the Seine-et-Bièvre, the Hauts-de-Seine, and the Plaine-St.-Denis, unless one of them is more elegantly called Versailles. Similar news of modernizing is that the French telephone exchanges, with their historic and informative names, such as Gounod and Chénier, will soon become mere numerals on the phone dial—408 for the composer of "Faust" and 253 for the revolutionary youth whose statue is in the Palais-Royal Garden, where he spoke on behalf of liberty. Even Balzac will become merely 225.

June 4

This coming Saturday will be the twentieth anniversary of the English, Canadian, and American landings on Omaha and Utah Beaches in the decisive invasion of France—by them and not by the French, by their generals and not by General de Gaulle. The French citizens, often wearied by or captious about de Gaulle's standoffish reaction to the Anglo-Saxons, seem solidly behind him in his refusal to participate in this anniversary on those bloody, historic sands of his native land where he was not invited to set foot.

December 16

This year's end finds President de Gaulle on a peculiar pinnacle of lonely leadership—lonelier by far than the isolated eminence he has always contrived for himself. In a little more than a twelvemonth, through the brutal loss of President Kennedy, the unexpected sacking of Premier Khrushchev, and the sequestration into old age of Sir Winston Churchill, de Gaulle has emerged as the unique familiar, powerful political and historical figure left on the Western world's governing scene. Turned seventy four last month, he apparently supports both time and the weight of his heavy office without any mortal weakening, and certainly with no abatement in his spiritual and intellectual concentration on the guiding of France always steeply upward, toward the floating status of international power. As encouragement, the miraculous fiscal prosperity of his Fifth Republic, now entering its seventh year, has reached new amplitudes. Today, France is one of the richest nations. In a discreet communiqué just issued by the Finance Ministry, it was made known that the country's official reserves have now reached five billion dollars, and that the gold and foreign *devises* in the vaults of the Banque de France are unequalled in Europe—except by the Bundesbank, in Bonn. Diplomatically and geographically, de Gaulle has had an especially expansive year in 1964, what with his state tour of Latin America, accompanied by Madame his wife and a suite of twenty, on which he astonished the South Americans by his bravura in making very brief speeches in Spanish—a tour de force that impressed even those compatriots back home who had become increasingly irritated by his highhanded governing methods and the accompanying Olympian manner.

In his native tongue, de Gaulle's speech last month in Strasbourg, on the twentieth anniversary of the liberation of the city from the Nazis, seems to have ranked as a real national event. It rated all over France as notable in his oratorical and thespian career—a speech still mentioned here in Paris as an exceptional example of his fine French language, his personal literary style, and his diplomatic skill. Certainly, de Gaulle's reference in it to his treaty of coöperation with West Germany after three centuries of enmity and his blunt declaration that the treaty had been aimed at "the construction of a European Europe" were intended to be heard across the Rhine by the Germans, and also across the ocean by the Americans. His distrust of the United States-sponsored Multilateral Nuclear Force (now being discussed at the NATO Council meeting in Paris) could be located behind his statement that in this push-button nuclear age there is no way "to insure the initial safeguarding of the old Continent, and, in consequence, to justify the Atlantic Alliance . . . except by the organization of a Europe that will be its own self, so as to defend itself"—and not merely to be rescued much later by others as part of their own salvation.

What de Gaulle did not mention at Strasbourg was the fact that in 1965 the West Germans will be holding their national elections. It is known that de Gaulle was outraged when the Bonn Minister of Foreign Affairs recently told his French counterpart, M. Couve de Murville, that the Federal Germans in their election propaganda will boldly contest the Oder-Neisse line, which was laid down in the Allied peace treaty as the German-Polish border. It is interesting to recall that in the third volume of de Gaulle's memoirs he recounts how in 1944, on a hasty and probably unwelcome visit to Stalin in Moscow, with the war not yet over and with him, as the French Resistance leader, still practically a nobody, he rejected Stalin's peace proposal that all Poland be sacrificed, and retired to his quarters in a dudgeon, to be awakened in the middle of the night with a message to return to Stalin, who then accepted de Gaulle's position. Another German irritant to de Gaulle has been their telling him that despite the Franco-German amity treaty, they have no intention of giving the go-by to their dear, generous old Uncle Sam. This semi-infidelity has been the major setback to de Gaulle's still precious project of a new Europe in which France—if she is to lead it and remain safely in first place—requires a half Germany as her second fiddle. Certain diplomats here believe that it is as an elderly historian expert even on modern history that de Gaulle has so well understood France's position in it—a relatively weak power playing the part of a first-class one.

France, too, will be holding her national elections in 1965, and nobody in his right mind here thinks that President de Gaulle will not be reëlected next November—if he runs for office. Since he will be eighty-two in 1972, when his new term would expire, the only serious curiosity now is about the succession, which would surely start operating before that date, and about exactly how Premier Pompidou, his present choice as his *dauphin,* might be legally inserted as the succeeding member of the Gaullist dynasty. The other, even greater question, looking ahead to the time when de Gaulle himself will no longer be in personal control, is how didactically reëducated the French Parliament will prove to have been by these last six enforced do-nothing years, during which the General, who did not trust it after its Chamber-wrecking activities in the two previous republics, practically put it out to grass, to ponder. The editor of a liberal-left Paris weekly, always in opposition to de Gaulle for his complete holding of power in his enormous hands—his thumbs looking as big as a pair of upright bananas in his famous gestures on television—has just admitted that no modern democracy, like France or Italy, has been able to function efficiently with a multi-party government. The final question is: Has the Fifth Republic's Parliament finally settled that fact firmly in scattered comprehension?

Actually, this may be the right moment to set down a kind of rough-and-ready approximation of the vital differences between the two preceding Republics and de Gaulle's, so as to obtain a year-end notion of where contemporary republican government in France has been, is, and may be going. As Anatole France said of the Third Republic (1871–1940), in which he lived and died, its main characteristic was *la facilité*. It was the easygoing *République des Camarades*—of pals—with financial and political scandals that enriched politicians and shook down governments and unfortunate petty investors like sere autumn leaves several times a year. Some of its nineteenth-century scandals were like bold, melodramatic, up-to-date farces—such as Mme. Présidente Hanau and her bank scandal, followed by her suicide in prison, where she slept on her own linen sheets and under her own mink coat. Then in the nineteen-thirties came the really alarming Stavisky scandal, with its government-backed provincial pawnshops and bogus pawned emeralds, which involved both France's Premier and his Cabinet and led to the notorious Place de la Concorde riot of February 6, 1934, when eighteen protesting citizens in the mob trying to assault the Parliament and the deputies were shot down and killed by the Garde Mobile, and the Paris climate smelled like the beginning of a Fascist civil war. In the Fourth Republic (1944–58), political morality became close to impeccable, despite the unbridled Ballets Roses scandal of the one-armed Speaker of Parliament, André Le Troquer, and the wine-trafficking scandals of a Socialist Minister. Few republics in our time anywhere have had a public servant so intelligently used as M. Jean Monnet, father of the European Common Market, though the Fourth Republic and its young technocrats had little praise in France for their farsighted faith in him. Many anti-Gaullists of the upper class still contend that it was the Fourth Republic that brought salvation and order to France's opaque finances, and should thus claim the credit for founding the base of the Fifth Republic's prosperity. Actually, the prosperity resulted more from the ending of the Algerian war, which the Fourth could neither win nor finish—the economy of no longer shedding blood and money in a colonial war. It is true that the technical handling of French finances had already improved under the Fourth's technocrats, but the improvements continued, with compounded results, in the Fifth. The great amelioration in politicos' brains from the Third Republic on lies in the fact that it was the graduates of the Ecole Normale who dimly shone as Ministerial lights in the old days, whereas in the Fourth and Fifth it has been brilliantly educated men from the Polytechnique, the severest scientific training ground in France. French higher education, starved as it is for decent classrooms, properly equipped laboratories, and decently paid faculties, has nevertheless managed to improve the quality of the students' intelligences, probably because they are still obstinately determined to achieve knowledge and culture, and study to obtain both. The whole level of economic knowledge has enormously risen since the Third Republic days, when France's economic teaching was rated by the English and Germans as among the most backward in Europe. The financial experts' level in the Fifth Republic is now high and sharp indeed, according to British bankers who had to deal with them in the recent loan, via the International

Monetary Fund, to tide over the Bank of England in the vulpine international speculators' attack on the staggering British pound.

What certain veteran foreign visitors, familiar with France during all three Republics, say of France today is that the most important fact now is not the age of its General de Gaulle but the increasing youth of its population—a refreshed situation apparently not known in this country for the last hundred years. It is they who will make the new France, whatever it turns out to be.

INDEX

Index